THE COMPLETE GUIDE TO Building Your Own TREE HOUSE

For Parents and Adults Who are Kids at Heart — With Companion CD-ROM

By Robert Miskimon

The Complete Guide to Building Your Own Tree House: *For Parents and Adults Who are Kids at Heart —With Companion CD-ROM*

Copyright © 2010 Atlantic Publishing Group, Inc.

1405 SW 6th Avenue • Ocala, Florida 34471 • 800-814-1132 • 352-622-1875–Fax

Web site: www.atlantic-pub.com • E-mail: sales@atlantic-pub.com

SAN Number: 268-1250 • Member American Library Association

ISBN-13: 978-1-60138-244-3 •   ISBN-10: 1-60138-244-8

Library of Congress Cataloging-in-Publication Data

Miskimon, Robert, 1943-

   The complete guide to building your own tree house : for parents and adults who are kids at heart-with companion CD-ROM / by Robert Miskimo.

       p. cm.

   Includes bibliographical references and index.

   ISBN-13: 978-1-60138-244-3 (alk. paper)

   ISBN-10: 1-60138-244-8 (alk. paper)

   1. Tree houses--Design and construction. I. Title.

   TH4885.M57 2009

   690'.89--dc22

                            2009033945

                    10 9 8 7 6 5 4 3 2

PROJECT MANAGER: Carrie Speight

COVER & INTERIOR DESIGN: Meg Buchner • megadesn@mchsi.com

PRODUCTION DESIGN: Jackie Miller • sullmill@charter.net

ASSISTANT EDITOR: Angela Pham • apham@atlantic-pub.com

Printed in the United States.

Acknowledgements

Many people helped to make this book a reality. I am particularly grateful to Michael Garnier of Out 'n' About Treesort, Pete Nelson of Seattle, Meghan Welch of Bethesda Lutheran Communities, Ray Cirino of Los Angeles, Barbara Butler of San Francisco, Luke Lukoskie of Vashon Island, Washington, and to all those who helped by providing me with their stories to be used as case studies. To my editor, Carrie Speight, and especially, to my grandmother, Mae Martyn Miskimon, who first encouraged me to become a writer.

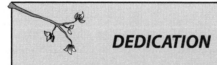

DEDICATION

*To Elizabeth, Mae, and Anina —
and to all children of all
ages everywhere.*

Table of Contents

Chapter 2: Planning and Design 21

Chapter 3: Tools and Materials 31

Chapter 4: Safety Concerns 41

Chapter 5: Building the Platform 51

Chapter 6: Doors, Walls, and Windows 63

Chapter 7: Raise High the Roof Beam 77

Chapter 8: Out 'n' About Treesort 87

Chapter 9: The Most Important Part: Getting the Kids Involved 93

Chapter 10: Access and Accessories 103

Chapter 11: Tree House Plans 115

Foreword

In the past few years, there has been a boom of interest in tree houses, but there are not many resources that show the average person how to build his or her own. This is where author Robert Miskimon and *The Complete Guide to Building Your Own Tree House: For Parents and Adults Who Are Kids at Heart* come in.

In an easy-to-understand style, Miskimon covers all the bases — from initial tree selection all the way to the final coat of paint. In between, nothing is left out. He covers safety, types of wood, fastener selection, and much more. That information alone is worth the price of this book.

The Complete Guide to Building Your Own Tree House is a book you will want to keep close at hand while building. You will want it for checking joist spans when planning the design; you will want it to check pilot hole sizes when drilling for lag placements; and you will also want it to check the recommended sizes of all the accessories your kids will want in their new abode. Do you know how far apart to space the rungs of a vertical ladder? You will after reading this book.

It is refreshing to find a tree house guide that focuses on the practical aspects of tree house construction. This is not a coffee table book — it's a router-table book.

I believe everyone should have the right to experience being cradled in the boughs of a tree. It has been the focus of my life these past few years, and even though we build about one tree house per week, to make that dream a reality, we really need an army of empowered individuals creating their own special hideaways. *The Complete Guide to Building Your Own Tree House* will aid that goal. I wholeheartedly encourage you to read it, apply it, and most of all, enjoy the fruits of your labor. Happy building!

Steven Chmielnicki
Artisan Tree & Treehouse, LLC
Bryn Mawr, PA
www.artisantrees.com

Introduction

A tree house, whether plain or fancy, is a castle in the sky, a flying ship that chases wild geese around the moon, a quiet refuge from the world, and a door to the realms of pure imagination for children of all ages — including adults. There is truly something magical about being up in the limbs of a tree, cradled in the arms of nature, safe from the marauding pirates below.

If you built a tree house with your parents as a kid, or if you are a parent and you want to build a tree house for your own children, you will find this book useful as you begin to contemplate the nuts and bolts of location, design, construction, and safety (among other basics).

Whether simple or intricate, your tree house will be a learning experience. If there is a young potential carpenter in your family, it is a great opportunity for him or her to watch, learn, and provide an extra pair of hands.

Of course, once the tree house gets started, you will probably want to include some fun extras, such as ladders, trap doors, skylights, zip lines, fireman's poles, and perches. You can even make your own "tree-a-phone" with two empty tin cans and a length of high-quality twine.

In this book, you will see that there is not just one way to build a tree house, and that these wonderful structures can be as individual as the personality and imagination of the builder. If you are an inexperienced builder, this book gives you basic concepts and tools that you can use in a variety of applications.

You will find words of wisdom from fellow tree house builders throughout the book in sections titled "case studies." You will find these at the end of each chapter and in the appendix. They offer stories of tree house building experiences, and will provide you with useful advice, keen insight, and entertainment.

For example, you will find the Case Study of professional builder Scott Daves, who first became interested in the construction trade by building his own tree house as a kid. Now he is an adult with a successful home-building business. He used his experience to keep up the tree house tradition by working with his son to build his first tree house. He demonstrates that building a tree house is a fun and rewarding way to spend family time together.

You will meet Barbara Butler, whose whimsical, cartoonish tree houses and playhouses for children have become immensely popular with celebrities as well as everyday folks. You will learn the story of Michael Garnier, the Oregon tree house trailblazer who won an eight-year battle with local authorities to get permits for his tree house resort and who is in the process of becoming an iconic figure in the tree house movement.

Los Angeles inventor Ray Cirino's experience is also included in this book. Cirino has built tree houses mostly out of objects he finds, such as recycled aircraft metals, driftwood, and other unusual materials. He has perfected a unique method of suspension involving the use of rubber rings around the trunk that minimize tree damage. His story, along with many others, is featured in this book to provide you with extensive knowledge of the process of building a tree house.

Along with these case studies, you will find valuable information in each chapter that will assist you along your way to creating your tree house. You will find techniques for roofing, flooring, and decorating. A special chapter just for kids will help you get your children involved in this process. You will also find a chapter that outlines some of the fun accessories you can add to your tree house to enhance your child's imagination and creativity.

This book will provide you with everything you need to know to build your tree house. You can find pictures of tree houses, as well as 15 tree house plans and a tool glossary complete with illustrations. Good luck on this adventure, and may you have as much fun making a tree house as you will playing in it.

When it comes to imagining and building your tree house, the sky is the limit.

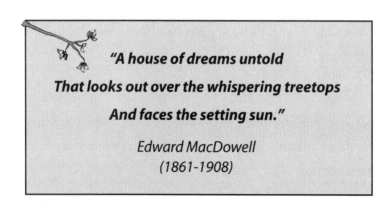

"A house of dreams untold

That looks out over the whispering treetops

And faces the setting sun."

Edward MacDowell
(1861-1908)

Getting Your Feet on the Ground

Location, Location, Location

Where you build your tree house is just as important as how you build it. Walk around your house and property and take note of the various trees — their species, size, age, and evident health. Older trees are generally bigger trees, which makes them more able to support a tree house. Make note of where these mature trees are located, keeping in mind the proximity of two or more large trees as support for your structure.

As you observe and assess the potential of various trees, remember to inspect the foundation and look for any signs of disease, such as damaged bark or fungus where the trunk contacts the ground. Trees that have compacted earth around their foundation are not the best trees for a tree house because the earth cuts off their nutrients. If a tree is too close to a paved driveway or busy footpath, it may have root damage that causes it to get sick and die. If you were to build a tree house here, it could result in a structural failure.

Inspect the trees for any large openings in the bark. Look for spots where the tree is healing from injuries to see whether the exposed wood is solid or rotting. If there is a question of whether the tree is hollow because of damage and rot, tap on the trunk with a 2- by 4-inch length of lumber. If there are changes in tone as you thump the trunk, it is a sign that there is hollowing inside.

Getting Started

Building a tree house is much like building any other house. The planning component is logical and linear; however, it is important to allow room for that creative impulse that gives your project its personality. With a tree house, you will be considering several factors at once. Some of these factors are:

- Looking for appropriate tree(s) to use
- Considering the height of the branches from the ground
- Figuring out what you will need to get into and out of the tree house
- Considering whether you want just a platform or a house

No matter what your desire, the tree you choose is the ultimate factor for determining what you can build in it.

Who Will Use the Tree House?

If your tree house will be a kids-only structure, you have to build it "kid's size," which means looking at it from two angles: from a "kid's-eye" view and a safety view.

Have you ever gone back to a play area of your childhood and wondered how everything had shrunk so much? To a 6-foot-tall man, a platform 6 feet from the ground is nothing — but to a 3-foot-tall child, it is like Mount Everest. Thus, a kids-only tree house does not have to be very high. Anywhere from 4 to 8 feet from the ground is quite sufficient, with 8 feet being probably the utmost limit. These heights are enough to give kids what they want (elevation) and what you want (relative safety).

Naturally, if the tree house is going to be used by both kids and adults, these same height considerations should be taken into account. A fall of more than 8 feet might not be much to an adult, depending on the landing, but to a kid, it can be deadly.

An adults-only tree house can be any desired height from the ground, so long as basic safety features like sturdy rails and steps are incorporated. If you plan to build a large tree house, it makes sense to consult with a structural engineer who can figure load capacity based on the size of the tree house, as well as an arborist who can give you advice on tree strength, growth, and sway potentials.

Choosing the Tree

Not too long ago, a tall palm tree on the grounds of a beachside hotel in Southern California suddenly lost its entire top branches. The top fell, killing a sunbather. Another common tree in Southern California, the eucalyptus, is notorious for dropping large limbs on parked cars and pedestrians with no warning. There does not have to be a wind or any other stress on the tree for it to drop its limbs and branches.

From this information, you can gather that two trees you probably do not want to use for a tree house are the palm and the eucalyptus. But there are a number of other trees unsuitable for tree houses as well. In general, unsuitable trees are those with shallow root systems, brittle branches, too many dead limbs and branches, or hollow trunks and limbs. Any tree with signs of rot or disease should be avoided. If in doubt, get the professional advice of a local arborist.

Good Species versus Bad Species

The table below is a guideline for potentially good and bad trees for your tree house. Reasons for deeming a tree unacceptable for a house include shallow root systems, wood that is too soft for lag screws, brittle branches, shorter lifespan, and unsuitable size and strength of their trunk and limbs. If you are going to build a tree house that is not supported entirely by the tree, you could perhaps get away with using one of the trees from the "bad" column as the tethering point for one corner of your platform, but be careful in your design and construction if you do.

TREE SPECIES			
Good		**Bad**	
• **Apple**	• **Ash**	• **Aspen**	• **Birch**
• **Beech**	• **Cedar**	• **Black Walnut**	• **Box Elder**
• **Cypress**	• **Douglas Fir**	• **Cottonwood**	• **Eucalyptus**
• **Hemlock**	• **Hickory**	• **Locust**	• **Lombardy Poplar**
• **Maple:** Sugar, Red		• **Maple:** Silver	• **Palm**
• **Oak:** White, Live Oak, Northern Red		• **Spruce**	• **Virginia Pine**
• **Pine:** Ponderosa, Eastern White, Sugar			
• **Spruce:** Colorado Blue, Black, Engelmann			

Size and Orientation: Trunk and Limbs

The trunk and limbs of your tree house tree must be large and strong enough to bear the stresses of the tree house. The more mature and bigger the tree, the less it will sway in the wind. A tree house that sways too much will be like the cradle in the classic lullaby: It will fall, taking baby and all. Trees are in a constant fight with gravity. If you build your tree house in a tree that is leaning very far to the right or left, you will be adding gravitationally induced stress to the roots, and the tree could topple. Here are some general guidelines for the trunks and limbs:

1. The trunk of a tree supporting a tree house should be at least 2 feet in diameter (width) at the base of the tree.

2. The trunk of a tree (or trees) that are not going to carry the entire load can be as small as, but no smaller than, 6 to 8 inches in diameter.

3. Limbs that will support any part of the platform should be at least 6 inches in diameter.

4. The trunk should be perfectly vertical. Reject a tree with a leaning trunk.

5. The limbs that will carry the platform load should be as parallel to the ground as possible. Although it might not seem to be the case, limbs that branch upward at an angle from the trunk are actually weaker than ones growing straight out at a 90-degree angle.

Now let us take a look at some pictures of trees and read an analysis about their structure.

Use these pictures as examples when looking for an appropriate tree for your tree house.

A tree that is leaning, such as this one, is not a good choice for a tree house. The weight of the tree house can add stress that can uproot the tree.

Wind damage to this tree with the resulting dead limbs makes it unsafe for a tree house. One could trim the tree so the two main trunks are 6 to 8 feet high and use them to anchor a platform.

Limbs that extend at or about 90 degrees from the trunk, as with this tree, are preferred for bearing the weight of a tree house platform.

A tree house platform could be constructed between the forks of this tree, so long as provision is made for wind-caused movement.

This tree meets the minimum size requirements, being straight and having a trunk of more than 2 feet in diameter at the ground.

Check the condition of the root structure of your potential tree house tree. It should be solid and healthy like the ones at left.

Unwanted Branches and Limbs

Few trees are absolutely perfect when it comes to accommodating a tree house design, and there might be dead branches that pose a potential hazard. Consequently, you might have to remove one or more limbs. If, for some reason you cannot do this yourself (for example, there might be a dangerous limb higher than your ladder can reach), call a professional tree trimmer. The cost is a lot less than a potential trip to the emergency room.

A Dozen Ways to Create Safety

Experts who study industrial accidents such as plane crashes and nuclear power plant meltdowns to prevent such mishaps in the future believe that safety is created anew each day. Safety is the byproduct of certain patterns of thoughts and behaviors that tend to reduce the probability of unintended consequences. For example, a surgeon who disinfects his or her hands before an operation greatly reduces the probability of the patient becoming infected.

Likewise, there are some suggestions for tree house builders that can help you foster safety as you work on your project:

1. Before you begin, carefully examine the tree and pick your climbing route.

2. Always have a rope or an accessible limb nearby in case you lose your balance while climbing up the tree.

3. With a handsaw or clippers, chop off any dead wood on the tree that could break off while you are climbing or working.

4. Use a tool belt fastened around your waist to carry tools, nails, and screws, and to keep your hands free for climbing.

5. Adhere to a regular maintenance schedule (see Chapter 4) for your tree house, checking for rot, loose wood, leaks, and other potential issues.

6. To raise and hold beams in place while installing them, put a rope and pulley high in the tree.

7. Be sure to prevent splinters and cuts by sanding any rough wood and removing old or rusty nails.

8. Encircle the base of the tree(s) with some kind of soft ground cover, such as bark or sawdust, to cushion anyone who might fall.

9. Unless you are a professional tree house builder, do not build too high in the tree; most tree houses built for children are 6 to 8 high.

10. Whenever you move the ladder, verify each time that it is level and stable.

11. If you use an extension ladder, be sure it faces in the correct direction before mounting it.

12. Provide extra reinforcement for the tree house's platform to keep it sturdy and firm despite weathering and hard use by children.

Because experience is often the best teacher, the following is a Case Study presenting the experience of a veteran tree house builder. This Case Study, along with the many throughout the book, will offer you keen insight into the tree house building experience.

Case Study: Tips on Building a Tree House From a Professional Amateur

CLASSIFIED CASE STUDIES TM
directly from the experts

By Konrad Kaletsch

Consumer Advocate

Saugerties, New York

Here are a few bits of wisdom I have acquired as an amateur builder on my fourth tree house:

Tree houses high in the sky are great, but they are hard on the people who build them. Building involves lots of bending and twisting actions — the higher up you are, the farther you fall.

Build as much on the ground as you can, and hoist up completed sections. This reduces your "air" time.

It matters how many trees you use and how high they are. Building off one trunk is trouble-free, but if the wind moves the tree, the whole tree house goes with it. If you build on two or more trunks, you have to consider the stresses the movement places on your hardware. It is counter-intuitive, but the closer to the ground you are with multiple-tree structures, the more important it is for your hardware to allow movement.

The many techniques for doing this must be evaluated based on the size and weight of the tree house. Typically, each board gets bolted to one trunk directly and then has a sliding device on the other tree trunk. Be consistent: One tree is the fixed tree, and the other is the slider.

Testing: As you build, more and more weight is added. Do your homework to build a structurally sound tree house. If you are unsure, consult with a professional builder, engineer, or architect. But no tree house can be declared safe without a test. One method is to place empty oil drums inside with an attached water hose. From a safe distance, turn on the tap and fill the drums. You are creating at least double what the occupancy weight will be. Consider also the weight of snow in northern locations. Let the drums sit for a week, then drain them remotely as well.

Well-positioned hoses in the drums will allow you to siphon out the water. If the weight collapses the tree house, you do not want to be in or under it.

Not all trees are suitable for tree houses. Try to use hard woods as much as possible in the construction of your tree house.

CHAPTER

Planning and Design

Research Zoning and Building Codes

It is worth mentioning that a little research into local regulations concerning tree houses before you start to build one can save you a lot of potential headaches down the road.

Fictional adventurer Robinson Crusoe probably could not get away with building his fascinating home today. Back then, in the days of Daniel Defoe's imagination, there were not building codes or planning and zoning laws. But, on the positive side, today's tree house builder has access to far superior building materials and a wealth of practical experience encoded in building regulations.

Once you have a rough idea of what and where you want to build, a call to your local planning and zoning department is a good idea. Some cities basically overlook tree houses as too small to spend their time on. Some locales have a hands-off attitude toward tree houses; others may require you to first get a building permit.

Some city and county governments view tree houses as temporary structures and impose limits on their overall size — usually between 100 to 120 square feet, with a height restriction of 10 to 12 feet. Because of the elastic nature of trees as living, changing organisms, it is difficult to define strength and stability standards, thus many jurisdictions do not. If this is the case where you live, following the guidelines in this book will help you to build a safe, structurally sound tree house that your kids will love.

Your local planning office may simply advise you to respect their setback requirements that define how close to your property line you can build, and not build in a conspicuous place where neighbors might object or even file a complaint. Talking to your neighbors about your tree house is diplomatic and practical; you should ask for their suggestions about location and visibility.

Respect Privacy

Respecting your neighbors' privacy is very important if you want to keep a friendly relationship with them. Be sensitive to placement of windows and decks so they do not provide intrusive visual access to the inside of neighbors' homes. Think of a location and style for your tree house that will have minimal impact on the viewshed, or visual surroundings, in the environment. It is a good idea to respect any height restrictions for backyard structures, which may be the same as for garages and tool sheds. Remember, after taking into account privacy and height requirements, avoid building in your front yard or anywhere else easily visible to the public. These considerations will ensure that you are less likely to get complaints from neighbors.

One sure way to get the attention of local officials is to include electricity and plumbing in your tree house. At that point, your dream project goes from being a temporary structure to a residence that is subject to all the restrictions and requirements of the local building code. Just ask any builder or contractor how much fun they have working through bureaucratic red tape, and you will see the wisdom of not including electricity or plumbing.

Nevertheless, as mentioned above, local building codes reflect many years of defining the best standards for building safely. It might be a good idea to read the local building code to check the standards for such elements as floor joist spans, railings, and support beams. Consider making your plan include consideration of local geological and meteorological issues.

Elements of Design

Although some people use a "design-as-you-build" approach to tree houses, this method is potentially more troublesome than using a thoughtful plan. The plan should take into account the factors that can make the difference between an improvised structure that may have to be modified or torn down, versus one in which potential problems are thought-out ahead of time.

Before designing your tree house, it is a good idea to consult an arborist (tree specialist) about such factors as tree health, sustainable weight loading, and stress caused by tree movement in the wind.

Important elements of tree house design include:

- **Platform.** This is the basic element that will make your tree house a reality. The shape and overall size of the platform should be compatible with a one-tree, two-tree, or three-tree design. In addition, building a tree house should be fun and imaginative, so building an odd-shaped or irregular platform is all right, provided it affords good support and weight distribution. If children will be the primary users of the tree house, a 6- to 8-foot elevation from the ground is a good idea for safety. See Chapter 5 for more information about platforms and a more thorough discussion of building the platforms.

- **Walls.** In addition to their aesthetic impact, walls provide a measure of support for the roof and can be solid and made of plywood or some other material, or latticed to allow breezes to flow through the house. Depending on how far your design goes to incorporate the shape and curve of surrounding tree(s), your walls can be circular — if surrounding a tree — square, rectangular, or irregular. Height of the walls need not be the 8 feet standard in regular homes. Six feet of headroom is usually enough for most kids, and maybe another ½ inch for adults.

- **Roof.** Your roof can be flat, slanted, conical, gabled, square, or just about any shape you can imagine, so long as it performs the necessary function of shielding you from the elements. It can be solid, or open with skylights. A roof can be permanent or removable and made of canvas or a plastic tarp. It can be covered with composite roofing materials, tin, or shingles, depending on your taste and pocketbook. If you are considering a skylight, make sure you figure on leaving room for the cutout in the roof.

- **Windows and doors.** A very simple door can be fashioned with a series of vertical wooden planks held together with a Z-shaped crossbeam, or that same door can be halved to make fun and interesting Dutch doors that open independently. Windows can be store-bought or customized to your tree house; you will need to create a frame for the window in the wall supports, sized to the scale of your house. Another possibility is to glue a single piece of Plexiglas™ into a wooden frame, held in place with wooden strips at the edges that will fit into the wall. It is highly recommended to use some form of window because closed spaces can become quite dark and moldy. This can cause wood rot with its attendant dangers. Sometimes found or discarded window frames can be well-used in your project, and you might run across a salvaged window frame in, say, the style of a porthole with hinged glass. In the world of tree houses, let your imagination run free as you visualize the various components and how they might all fit together.

- **Support posts.** When the available tree(s) by themselves are not positioned to support a tree house, you may need to build support beams, or when the platform/deck extends far enough away from the main structure so that additional support is needed. These can be simple 4-inch beams anchored by a concrete pier (a large piece of concrete that can serve many purposes as a foundation for a structure), or resting in the ground and encased in poured concrete. Using support beams necessarily means your tree house will be lower to the ground to minimize the effects of tree movement. A good rule of thumb is to keep support beams to within 10 to 12 percent of the tree's total height. So if you are building in a 50-foot tree, the beams should be no more than 5 feet high.

- **Deck.** Because a deck is a universal favorite among tree house builders and dwellers, it is wise to consider early in your planning what size and shape deck you would like. A deck not only puts you outside in nice weather, but adds to the overall usable space in the tree house. Plus, a deck is the logical place to anchor the high end of a zip line or rope ladder.

Drawing the Plans

At this point, your mind may be bursting with ideas and questions. Before you take up hammer and nail, this is a good time to take a deep breath and write down what you are trying to achieve. A plan, however rough, can help you to spot potential trouble areas long before you are high up in a tree and cursing yourself for not thinking of something important, like the fact that your platform is not level.

Put a tape measure around the trunk(s) of the tree(s) you plan to use in constructing the tree house. Then sketch out in pencil the shape and dimensions of the platform and how it will be attached to the tree(s). This can be a simple, ruler-drawn set of lines superimposed on circles representing the tree trunks.

Another way to do this is to print out a large digital image of the actual trees that will support the structure, cover it with thin tracing paper, and sketch the outlines of the house directly over the photo so you can gain a better idea of size and proportions. Or, if you are fortunate enough to have access to a computer-assisted design software program (CAD) — such as architects and builders use — you can produce a detailed set of plans in only a few minutes.

A comprehensive set of plans that show how people would approach the tree house, any surrounding structures or bushes, and the general direction of interesting views can also be helpful. Draw sketches of the structure from four different angles, working your way visually around the tree house. These can be designated as "Elevation 1, 2, etc." In these sketches, you can also show the main branches of the supporting tree(s) from different heights down to the ground. These can be measured with a yardstick attached to a long stick to gauge the elevation of the branches.

From these initial elevation sketches, you can accurately draw where your platform should be built. Then you can proceed to sketch the platform from above and below, taking into account how major and minor branches of the tree will relate to the foundation. As you make these elevation drawings, you can go out to the actual trees to check your measurements against the plans. This is a good time to sketch in whatever form of access you want — stairs, a rope ladder, whatever — and get a clear picture of position and height of this feature.

How Many Trees?

To some extent, this question will be answered by the arrangement of trees in the location where you plan to build. In general, a single tree simplifies matters because the entire tree house structure is anchored to a single support — unless you decide to add supporting posts.

On the other hand, using two or more trees can allow a better distribution of weight and a somewhat larger structure. In regard to weight distribution, it is important to remember that, when building a tree house, you are superimposing a heavy mechanical structure onto a living, growing organism. Through millennia of evolution, trees have developed ways of coping with stresses of various kinds, including those caused by tree house building. Because the tree(s) provide support for your tree house, there are some important things to know about them:

- Damage to the bark can cause infections, just as a cut on your finger can become infected. The infection can result from airborne bacteria and fungi and cause localized rot and eventual death of the tree. Thus, the less damage to the bark, the better the health of the tree is likely to be. Tree growth enlarges the diameter of the trunk and can potentially damage the tree house unless you allow sufficient room for expansion. It would be wise to consult an arborist to estimate how much growth to expect.

- Cutting chunks out of the tree or branches to install supports can do lethal damage to the tree. Instead of using nails and screws to attach supports, use a single, large lag bolt (a heavy bolt used to support heavy structures) into a snug pilot hole or small hole that serves as a place to secure a larger bolt. This reduces the number of holes and eliminates compartmentalization, which is the process trees use to heal injuries. The tree forms a barrier (compartment) around the damaged area, seals it off, and stops sending nutrients to the area. The tree then continues to grow around the damaged spot. By using a single, large bolt instead of several bolts placed closely together, the tree heals the injury over time, and both tree and support bolt remain strong, without infection.

- Trees will grow to compensate for an altered distribution of weight by building a stronger root system to compensate for the stress. During this period of a few years while the tree is adjusting, it can be more vulnerable to storm damage. Achieving equitable weight distribution by building a circular structure around a single tree, or spreading the weight load over several trees, helps the tree(s) to cope with the demands placed upon them.

Four Corners or Three

With the additional support that two or more trees can provide for your structure, the design becomes important. If there are three trees that would seem suitable for supporting the tree house, you can easily build a three-cornered house. Or, with the addition of support beams, a four-cornered tree house is possible. There really is no limit to the number of supporting trees you can use, as in the case of a famous, huge tree house in Alnwick Garden in

Northumberland County, United Kingdom, built over 6,000 square feet.

In fact, you need not be limited by three- or four-cornered designs. Why not use a rhomboid or hexagonal shape, or something more organic in shape?

One key to using two or more trees is the use of flexible joints that permit the tree to move in the wind without tearing the tree house apart. Typically, one end of a support beam is permanently attached to the tree with a large, single bolt. A lag bolt should be at least 8 inches long and ¾ inches diameter. The other end of the beam allows the tree house to move in response to movement of the tree.

In designing a flexible support system, it is possible to allow the floor joist system to ride as an independent unit on the main supports, which permits additional flexibility. In this scenario, the joist system is created as a single piece attached at one end with runners and guides that maintain the correct position and integrity of the joists.

The two primary means of flexible support are metal brackets and cables. When mounting the support beams with brackets, allow enough extra length of support beam so it can slide freely within the bracket with no danger of sliding out of the bracket. Carefully measure the distance between the trees that the beam will span on a calm day when there is no wind. Mark on the beam where the stationary end will be bolted to the tree, and where the mid-point of the sliding bracket will be.

Then, screw a strip of steel ¼-inch thick and measuring 2 by 12 inches to the beam, centered on the mid-point. This metal bearing surface protects the beam from wearing away on the surface of the bracket. Brackets can be custom-made in a sheet metal shop for your individual tree house, and should be made of ¼-inch steel or thicker, welded together.

Next, mount the metal bracket in the tree with the realization that the actual surface of the floor will be about 8 inches above the surface of the bracket. Put the beam inside the bracket so that the metal sliding strip is in the right place; use a level to position the other end of the beam at the correct height, and temporarily fix it with a screw. Then drill a pilot hole in the correct place to affix the bracket. Pilot holes should be just slightly smaller than the diameter of the bolt for a tight fixture.

Cables as an Alternative to Sliding Beams

Using cables to support your tree house permits a wider range of motion in all directions, instead of a single plane, as is the case with brackets. There are a wide variety of cable types, as well as fittings that can be used for tree house construction. Steel cables also can be used to augment the strength of a long beam and are generally less expensive than using brackets.

Cables are appropriate in cases where a part of the platform extends too far from the tree to use wooden knee braces effectively, or when confronted with a curved or angled trunk or branch that prevents making a secure beam attachment using a bolt.

Cables should be attached to the tree with eyebolts. Do not wrap the cable around any part of the tree because it will wear off the bark and injure the tree. Be sure that the eyebolt enters the tree at a right angle to the cable for a secure and lasting fit. A turnbuckle enables you to set the cable tension after you place the beam and to adjust it periodically as needed. Do not allow any moving parts to come in contact with the tree to avoid damage to the bark. The supported beam should rest 6 to 8 inches away from the tree. You can also place a small section of wood on the tree to protect it against cable movement.

Design and Building Tips

- **Use new wood for support beams.** These are probably the most crucial structural and safety components of your tree house. They must be of good quality wood and adequate size to suspend the weight of your structure. You can ask at the lumberyard, but a rule of thumb in terms of size would be a 6- by 2-inch beam for spans between tress up to 3 feet, and an 8- by 2-inch beam for distances between 3 and 7 feet. Avoid knotty pine because of its reduced load-bearing capacity.

- **Carefully check recycled wood.** Often, it is the interesting and odd-shaped pieces of found wood that give a tree house its own personality and appearance. But when it comes to recycling old pieces of wood for structural members, you should carefully examine them before usage. If it is a beam, scrutinize it for any evidence of decay, rot, insect damage, and cracks. If uncertain about the quality of the wood, ask someone familiar with lumber to check it for you.

- **Avoid bolting support beams directly between tree trunks.** Even at lower levels, large trees can experience substantial movement in strong winds, causing the bolts to snap. The better way to attach beams is by using a flexible joint at one end to allow the structure slide, rather than shear, in that situation.

- **Use large bolts, not nails, for support.** A single, large lag bolt of ¾- to 1-inch diameter and length of 8 inches at each attachment to a tree not only causes less damage to the tree, but provides more structural support. Check support ratings for the lag bolts, and overbuild every joint to support three times the estimated weight of the structure with people on it.

- **Do not tie ropes, straps, or cables around the tree.** These will eventually strangle the tree and cause it to die. To allow for tree growth, provide a 2-inch gap around the tree if it will pass through the floor, and a 3-inch space it if passes through the roof. Observe the behavior of similar trees in windstorms to get an idea of how much sway you need to compensate for.

- **From both a safety and convenience standpoint,** it is best to build your tree house in sections on the ground and hoist them into position. This is a point at which you can involve the kids in hoisting built pieces up using a rope and pulley.

- **Diagonal braces, attached at the bottom to the tree and at the top to the platform, provide the greatest amount of support possible**. That is because of the inherent structural strength of a triangle as compared, say, with a square or rectangle. When you have aligned the beam and brace over each other, make a notch on the underside of the beam using a jigsaw, place the beam in the notch, and screw or bolt the beam to the brace.

Tree Size Matters

If you are planning a medium-sized tree house of around 100 square feet, look for mature trees with strong branches. Generally, the bigger the tree house, the bigger the tree needs to be for stability. If you want to build your structure using one tree for support, find a tree that measures at least 5 feet in circumference at its base. Main support limbs that anchor the corners of the house ought to be at least 6 inches in diameter.

The fact that trees move in the wind is one important structural issue often overlooked. If your tree house will use only a single tree, pick a large sturdy one and watch to see how it behaves in high winds. It is important to build the playhouse so the supporting tree is at the center for even weight distribution and less movement. If your architectural masterpiece will rest on two or more trees, it is advisable to use flexible joints in the platform. See Chapter 5 for more information.

At this stage of the project, it is a good idea to involve the most important consumers of the finished product — your children. Encourage them to help pick out the tree(s) that will be used for support in the best location, and ask for their ideas on how they want the tree house to look. You might also ask them to find any pictures of tree houses and put them in a pile for review and discussion in a family meeting. These meetings are also a good time to discuss safety issues with your children and to stress the need for constant awareness of potential hazards in building and using the new structure.

Now that you have done the preparatory work of scoping out potential sites and trees, as well as the legal and social impacts of your tree house project, you are ready to get started with a trip to the hardware or lumber store.

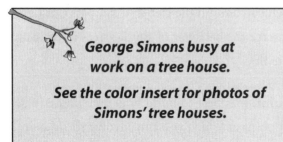

George Simons busy at work on a tree house.

See the color insert for photos of Simons' tree houses.

Case Study: Living Happily Off-the-Grid in the Trees of Maine

George Simons

Chef/food stylist

For some people, building a tree house easily morphs from a hobby to an obsession. Food stylist George Simons, being a level-headed sort of fellow, might be loathe to describe his embarkation on a fourth tree house as obsessive, but he does state emphatically that it is loads of fun.

George and his wife, Sue, had lived in Boston for a few years while they were in college, but George was not crazy about urban living.

"I grew up in Detroit and was in the Eagle Scouts," he said. "I had a lot of experiences camping outdoors, and I got antsy to get out in the woods. My wife is a city person and was reluctant at first, but after 10 years of looking, we found a nice piece of property deep in the Maine woods near a trout stream."

Their 40-acre property is in the foothills of the White Mountains, in an unincorporated area at about 900 feet of elevation from the valley floor. The property they purchased 15 years ago was affordable because it was not considered buildable land by the local authorities.

But before he became a Cordon Bleu chef and food stylist for print and television in Boston, George had served an apprenticeship with a master woodcarver from Oberamgerau, Germany, and he felt comfortable working with wood — although his construction experience was minimal, apart from building a tree house when he was a child.

Initially, the family used their property for camping excursions, but his "wife was a little nervous about being directly on the ground," George said. So he built an 8- by 12-foot triangular platform using three 12-inch-thick trees of oak and beech at an elevation of 15 feet.

Although he used a floating platform design, it really was not necessary because wind in that location is minimal, and eventually, the trees grew around the platform attachments through the process of inclusion. The platform supported an 8- by 12-foot structure that was tiny but serviceable when their two children were small, George said.

First tree house comfy, but confining

"I built right up the edge of the platform and ended up extending the deck by 6 feet to give us more space," he said. The bottom floor included a ladder up to the second-floor sleeping loft, and there was a mini-kitchen with a double-burner propane stove and propane heater.

For water, he used an RV pump to collect melted water from an old tin icebox that holds 3-gallon ice chunks. The ice keeps their food cold while melting.

"It was cozy but small," George now says of tree house No. 1. Five years later, George and Sue built a somewhat larger tree house with basically the same design as the first in a clearing about 50 feet away. Its basic dimensions are 12 by 12 feet with 8-foot ceilings, and it sits on a hillside. Plus, it has insulation and a wood-burning stove for those cold Maine winters. Access is by a spiral staircase around a tree and a bridge to the house.

"It has a 4- by 8-foot bump at the front that extends out from the platform to give more room," George said. The roof is designed so that the beam lines are staggered to allow installation of glass at the roofline. The basic platform is 8 feet wide, but George added a 2-foot "walk around" so he could easily complete construction. The second tree house was elevated 20 to 25 feet above a sloping hillside. Inside are a queen-sized bed and a night bathroom.

Unlike the first tree house, the second moved a lot in the wind, perhaps because of its higher elevation and greater exposure of the supporting trees to weather.

Simons describes the third tree house as a "ground house," even though it is suspended on three sides by oak and maple trees. The 6-foot porch deck creaks in the wind when it is supporting the trees' sway in the wind, but the house has been proved sturdy, George said. It is covered with board and batten siding, and has large, double-paned sliding glass doors in front.

The structure is 12 by 12 feet, with a second story. The whole thing rests on a platform that George describes as "monstrous." Currently, the first floor is unfinished and serves as a kind of office, while the second floor is paneled with tongue-in-groove wood.

Fourth tree house uses timber framing

A fourth tree house is a hexagon with wings, which George is building using timber framing, a method of construction that uses no nails but joins beams, walls, roofs, and flooring with carefully cut notches that hold everything together, "kind of like Tinker Toys," Sue explained.

"It will have a sleep tower, a kitchen, and a large eating area," Sue said. "The idea is to accommodate guests and larger groups."

Although their children were reaching adolescence by the time George completed his second tree house and were more interested in being with their friends than in a tree house in the woods, now that they are older, they like to visit on college holidays and use the tree houses as a place to stay when they go skiing in the winter.

As far as safety is concerned, George does complete maintenance checks of all the structures twice yearly but has not yet encountered any major problems, he says. By building his roofs with a large overhang, there is nowhere for rainwater to accumulate and cause rot. "Part of the beauty of this type of building is that the wood is not in contact with metal," George said.

Why does he keep building more tree houses?

"Because it is absolutely fun, and it is supposed to be fun," he said.

3

Tools and Materials

While it may be a poor workman who blames his tools, as the old saying goes, it is also the case that a good workman or woman chooses the right tools for the job with forethought and careful planning. Here are some suggestions for tools you will need in building your tree house.

Basic Tool Kit

Building a tree house calls for tools that cut wood, put pieces of wood together, and affix those pieces to the tree. With that in mind, the basic set of tools for a simple tree house with or without a structure on the platform calls for the following:

- Claw hammer (a 13-ounce hammer is best)

- Saw (crosscut handsaw and/or power circular saw)

- Level

- Screwdrivers (No. 2 and No. 3 Phillips head)

- Wrenches (open end, box end, and ratchet-and-socket)

- Drill and drill bits (a cordless drill is best)

- Tape measure (any length 10 feet or longer will do)

- Square(s)

- Pliers

- Wood-shaping tools, such as rasps

- Safety equipment: goggles, gloves, dust masks

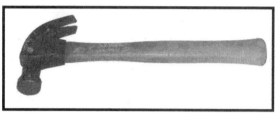

Claw hammer or framing hammer.
A weight of 13 ounces is minimum.

Levels. *Whatever the orientation of the tree's limbs, your platform has to be level with the ground. A regular carpenter's level should be sufficient for most needs.*

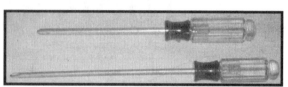

Screwdrivers. *A No. 2 and No. 3 screwdriver will be needed for deck screws.*

Saws. *A power circular saw will make quick work of cutting both sheet goods and dimensional lumber, but there might be circumstances, especially above the ground, when a handsaw will be a better choice.*

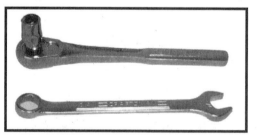

Wrenches. *To screw lag screws through lumber and into the tree, you are going to need at least an open-end or combination open-end and closed-end wrench. A socket wrench makes the job quicker once the lag screw is started.*

Retractable tape measure. *Best suited for building a tree house.*

Drills and drill bits. *A cordless drill is best for tree house building because you do not have to worry about running extension cords or getting power cords wrapped up around tree limbs and your own limbs. If possible, use two drills: one for drilling and one for driving screws. The specialty fitting at the bottom of this picture holds a drill bit in one end and a driver bit in the other so you can drill your pilot hole, then just reverse the bit to screw the screw into place.*

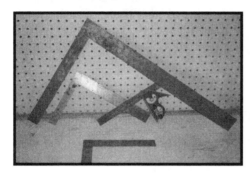

Squares. *At the minimum, you will need a framing square to be sure your right angles are really right angles.*

Advanced Tool Kit

If you are going to get fancy with your tree house, or if you just want to make some aspects of its construction a little easier, you will want to consider having some or all of the following in your tool arsenal:

- Miter box and saw (or compound radial-arm saw)
- Reciprocating saw
- Table (contractor's) saw
- Japanese flush saw
- Keyhole saw
- Sabre saw (also known as a hand-held jigsaw)
- Hacksaw
- Framing hammer

Miter box and backsaw or radial arm saw.

- Hand sledge
- Squares
- Wallboard square
- Framing square
- "Speed" square
- T-square
- Power sanders (orbital and/or belt)
- Hatchets
- Wood-chopping hatchet
- Roofing hammer/hatchet
- Pry bars (crow bar and/or small pry bar with nail slots)
- Auger bits (for drilling long holes in the tree or timbers)
- Post level
- Line level
- Pocket level
- Regular carpenter's level
- Framer's level
- Torpedo level
- Chalk line
- Chalk (stick)
- Duct tape
- Bolt cutter
- Cable cutter
- Router and router bits
- Router table
- Bar clamps
- C-clamps
- Pipe clamps
- Tools for post supports
- Shovel
- Posthole digger or auger
- Post level and/or plumb bob

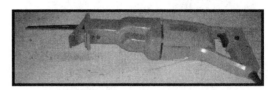

Reciprocating saw.

Pry bars.

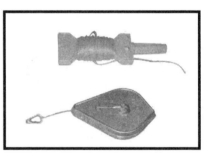

Chalk line and level line.

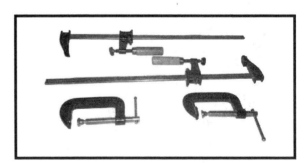

Clamps. Use clamps for holding boards in place when you are sawing them and when you are fastening them to one another or the tree.

All the tools listed here can be found at your local hardware store. In the end, the tools you use will depend on whether you already have them or if you want to spend the money on them, how well you know how to use them, and what you are expecting to build. If you want a really fancy tree house, you might want to seriously consider having a professional carpenter or contractor do the work; it could, in the long run, cost less in time, money, and headaches.

Hardware

At a minimum, you are going to need nails, screws, and lag bolts (also often called lag screws). For ease of joining floor joists and rafters, you will want to get some joist hangers. To accommodate the movement of the tree in the wind, you are going to need hardware that lets the tree move without harming your tree house or the tree. Important elements of tree house design include:

NAILS

When it comes to your basic structure, nails should be used only when necessary to tack a couple of pieces of wood to each other until you can screw them together. Nails are hard to remove if you have to, and they work themselves loose over time. Screws are easier to remove, and they do not work loose. This is a good rule of thumb for any exterior structure, including birdhouses.

SCREWS

Use exterior-grade screws (such as deck screws) that will not rust or corrode. Check the box label carefully — some exterior screws will rust, and the label should tell you that. The length of screw you use will depend on where you are putting them. In general, you should use a screw that will extend at least 2 inches into the last piece of wood that receives the screw. Do not use screws for fastening anything to the tree. Screws used with pressure-treated lumber must be the hot-dipped galvanized type, not just galvanized or coated with a polymer. The copper compounds used in pressure-treated lumber will eventually eat up any other type of screw or bolt. Again, the label on the box will tell you whether the screws inside it are rated for use with pressure-treated wood.

JOIST HANGERS

Just as with screws, any hanger used with pressure-treated wood must be especially galvanized for use with pressure-treated wood, and for the same reason. The bar-code label on the joist should indicate whether the joist is good for that purpose.

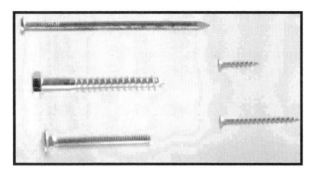

Assortment of nails and screws. *Hardware for your tree house will include common nails (top left), carriage bolts (bottom left), and deck screws (top and bottom right).*

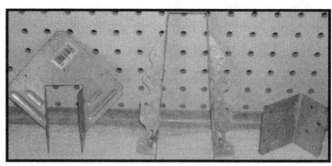

Joist hangers. *If you use hangers of any type, be sure they are galvanized. If the hangers are to be used for pressure-treated lumber, be sure they are hot-dipped and/or triple-galvanized.*

Screws. *Read the box to be sure the screws you buy are intended for the type of wood they will be used for. The screws in the top box are for cedar and pine, whereas the screws in the bottom box are especially coated for pressure-treated lumber.*

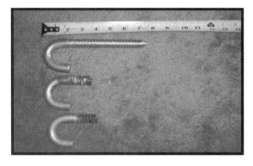

Failed arborist screws. *To prevent metal failure from fatigue, always use hardware at least ½ inch in diameter at the shank (not at the threads).*

Some specialty hardware (top to bottom): Through bolts, specially made support rods, Garnier limbs, and eye bolts. *Note that the eye part of the eye bolt shown is cast as one piece, not bent to form the "eye." Never use bent eye bolts, as they do not have sufficient load-bearing strength.*

HARDWARE ALLOWING MOVEMENT

You can get specially designed hardware that lets the tree sway in the wind without twisting your tree house into toothpicks. You can also make accommodation for such movement without special hardware.

Wood

Always go to the lumberyard and choose each piece of lumber yourself. It may be easier to pick up the phone and have "X" number of 2x4s and such delivered to your door, but when you do so, you will invariably end up with some lumber that is twisted, warped, bowed, and split. And you will be paying for lumber you cannot use.

Pick up each piece of lumber and sight down its length. If it is not straight in all respects, do not buy it. Also reject any piece of lumber with splits and with full or partial knots at its edges. Either of these defects decreases the strength of the lumber.

DIMENSIONAL LUMBER

When you are designing your tree house, you have to keep in mind the actual size of the lumber you are going to use. Most of us know that a 2x4 does not really measure 2 inches by 4 inches; it measures 1 ½ inches by 3 ½ inches. The following table shows the difference between "nominal size" (what we call a piece of dimensional lumber) and actual size.

Thickness x Width [inch]	
Nominal Size	**Actual Size**
2x2"	1-½ x 1-½"
2x4"	1-½ x 3-½"
2x6"	1-½ x 5-½"
2x8"	1-½ x 7-¼"*
2x10"	1-½ x 9-¼"*
1x10"	¾ x 9-¼"*
1x2"	¾ x 1-½"
1x4"	¾ x 3-½"
1x6"	¾ x 5-½"
1x8"	¾ x 7-¼" *
Note there is a ¾-inch difference in the width measurement, not a ½-inch difference.	

PLYWOOD

Although you can get plywood in sheets measuring 4 by 9 feet and 4 by 10 feet, and even larger, the most common size is 4 by 8 feet, which is what you will find at your neighborhood lumberyard. The widths from which you may choose are usually ⅛ inch, ¼ inch, ⅜ inch, ½ inch, ⅝ inch, ¾ inch, and 1 ⅛ inches. In general, you do not want to use anything less than ¾ inches thick if you are going to walk or put any significant weight on it. For sheathing purposes, such as for walls and roofs, the width you use will depend on its particular purpose.

GRADES OF PLYWOOD

Plywood is graded according to several factors, including the type of wood used for the outside layers, its allowed weather exposure, and the type of glue holding it all together.

On the grading stamp attached to a sheet of plywood, you will see that the first part indicates the relative defectiveness of the face and back veneers. Grade A is the best (although it is not always perfect), with the fewest defects, and Grade D has the worst defects. Therefore, a sheet of plywood graded A-C means the "face" is relatively defect-free, while the "back" has some defects.

- Grade A allows up to 18 defects, neatly repaired parallel with the grain.

- Grade B allows repairs and knots up to 1 inch, with some minor splits deemed acceptable.

- Grade C allows knots up to 1 ½ inch in diameter and knotholes up to 1 inch in diameter. Defects corrected by sanding are also allowed.

- Grade D allows knots and knotholes up to 2 ½ inches across the grain. Some splits are also allowed.

The particular face- and back-grade you choose is entirely up to you, but the one grading factor you should always keep in mind is whether the plywood is rated for exterior use. Always choose exterior grade.

Case Study: Gorilla Glue™ Holds This Tree House Together

Marcel Valliere
B-4/After Renovation Design
Rockland, Maine

Can you imagine building a large tree house on the steep slopes of coastal Maine overlooking the ocean — using Gorilla Glue?

Architectural designer Marcel Valliere of Rockland, Maine, could — and did. A precocious builder of tree houses who started when he was knee-high to a crossbeam, Valliere conspired with his wife, Jill, an interior designer, to undertake the project.

They bought a piece of property in Rockland while they were living and working in Boston, where Marcel attended the Massachusetts School of Design. About five years ago, he suggested they build a tree house.

"We are both whimsical," Valliere said. "We are our own children."

When he designed the structure, he was working for a design company in Boston that had a computer numerical cutting machine (CNC), a device that enables the designer to render the design accurately on software that directs the saw to make exact cuts. Using the CNC, Valliere designed a plywood truss system for the tree house platform made of ½- inch thick plywood for the exterior and 2- by 3-inch studs glued along the outside of the structure.

"We went through about a gallon of Gorilla Glue," Valliere says.

The 430-square foot structure is 16 by 18 feet, enclosed on two red oak trees. At each end of the square building, there are two decks that measure 16 by 8 feet. One end of the tree house looks out to the ocean, about a ½ -mile away, and the other is an entry accessible by a 16-foot walkway. The tree house is 14 feet off the ground.

"When you are on top, you are actually 30 feet off the ground," he says.

After building the platform, Valliere then mounted a 32-foot spine between the two trees, composed of three layers of 2- by 2-inch beams joined with Gorilla Glue and screws. In this manner, he fashioned two beams that rest in 1-inch-thick, stainless-steel brackets, secured to the trees with 36-inch steel pins. The platform, which rests on the two beams, slides whenever there is wind, which is often.

"We get extraordinarily high winds that come in off the ocean," Valliere said. "Other than the extensive use of glue, it is a typical square 2- by 4-inch construction that uses lots of recycled windows and other parts. We have a hip roof with a gabled dormer on the top side."

Valliere outfitted the tree house with a composting toilet and electrical power by running cords from the main house. He did not apply for any permits from the City of Rockland before building. "We took the attitude that it's easier to ask for forgiveness than permission," he says. The couple now uses the tree house as a guest cottage for friends, and charges no fees.

When city officials did inspect the structure, they decided to treat it as a garage and did not demand that he tear it down or modify it — although the structure did increase his property taxes. Finding insurance for the tree house was a challenge, but Valliere says he was finally able to get it covered under a homeowner's policy with Allstate.

Please see the color
insert for photos of
Valliere's tree house.

Safety Concerns

It is an obvious truism, but worth repeating: It is always best to think about safety prospectively rather than retrospectively. In other words, an ounce of prevention is worth a ton of cure, especially when building a tree house high in the treetops where falling objects may injure those on the ground, and workers themselves are exposed to hazards associated with power tools and falls.

Ladders

When building a tree house, you are going to have to be off the ground in one way or another, on a ladder or some other means of gaining height. Whether you use a short step stool or a three-section extension ladder, keep in mind that gravity is both friend and foe.

Steps to Ladder Safety

Nothing could be more discouraging at the outset than an avoidable construction accident, such as falling off a ladder or dropping a heavy tool near or on any bystanders below. Avoid such misfortune by following some simple safety rules:

- Have a crystal-clear idea of what your job is before you ascend.

- Do not leave a ladder up when you are not using it.

- If using a stepladder, position it exactly where you want it. If the ground is soft or irregular, use planks or odd-shaped pieces of wood to support the legs.

- Whenever possible, work on part of your tree house while on the ground and later hoist into place.

- It is dangerous to lean to one side or the other on a ladder. If something is just out of reach, descend and reposition the ladder.

- Use pulleys to hoist tools and materials to the building site.

- Insist that children and adults wear hardhats.

- Mark the perimeter of the construction zone with yellow tape to keep passers-by from wandering into the danger zone.

- No kids or visitors under the tree during construction.

- Do not use ladder rungs nailed to the tree in place of a ladder when building.

- Always climb a ladder with free hands; use a tool belt or pulley system to raise items to the work site.

- Work with a safety hook.

- Slow down. Move and work slowly, so the ladder does not begin to shift.

- Never allow two people on the ladder at the same time.

- Have a first aid kit nearby.

- Have a list of emergency numbers in your wallet and keep a cell phone handy.

- Use prescription or safety glasses for eye protection.

- Do not work alone while on a ladder.

- Do not leave power tools unattended.

- When you shop for a stepladder or regular ladder, it is worth a few extra dollars to get the highest quality.

- Be aware: Extension ladders, when extended, are high enough to come in contact with overhead power lines.

- Familiarize yourself with placement angles and weight limits for your ladder that are printed on the sides of the ladder.

- Do not use a ladder if your shoes are muddy.

- Always have three limbs (two feet and one hand, for example) in contact with the ladder rungs.

- Ascend and descend the ladder facing inwardly.

- Resist the temptation to jump down the last few rungs.

- Whenever you descend, keep an eye on the ground so you are not surprised and thrown off-balance when your feet hit the ground.

- To avoid clothing snags, place large rubber bands around the cuffs of pants and shirt.

- Be sure to have all the tools you will need when you ascend, so you do not have to go down and up again.

- Compress extension ladders for balance before moving them.

- If you want to pick up a long ladder from the ground, it is best to go to the top end, lift it, and slowly walk the ladder to a vertical position. It is easier and safer to position the ladder this way.

- Use all tools only for the jobs for which they are intended, and avoid using your ladder as a scaffold plank.

- Print your name and phone number on the side of your ladder and be judicious in lending it. Expect some damage to your ladder if you loan it to someone.

- Wasp stings are nasty and could cause you to fall from the ladder. Check the hollow rungs on your aluminum ladders for nests of any stinging insects.

- Stop working when you become fatigued; it is safer than pushing yourself and making a big mistake.

- Always make sure that the footing of your ladder is level and that it will not sink into the ground on one side or the other. Use boards, bricks, and paving stones — whatever you have at hand — to ensure this stability.

- Place your ladder so it is about 30 degrees from the vertical. A ladder that is too straight up and down will easily tip backward when you are at or near its top; a ladder that has too shallow of an angle can slip forward.

- Tie the top of the ladder to the tree or some other support when you can.

Gloves

Gloves have some important purposes other than protecting your hands from cuts and abrasions.

HANDLING PRESSURE-TREATED LUMBER

Although the copper compounds impregnated in pressure-treated wood have not been shown to be a health hazard, why take the chance? Arsenic compounds are still used in treating lumber that is meant to be in contact with water. If you are using that type of wood, gloves are highly recommended by the makers of pressure-treated lumber.

GRIP

You will want to wear good-quality gloves to improve your grip on tools and materials, to protect those on the ground as well as yourself. A good-quality leather glove helps you hold onto wood and tools so they do not slip from your grasp.

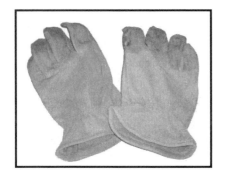

Gloves are a necessity for protecting you against the chemicals in pressure-treated lumber, and, as you can see from the condition of these gloves, they can protect your hands from other types of harm, like splinters.

Head Protection

Considering the people on the ground, you should invest in a hard hat for every person involved in helping you build the tree house. If you drop the socket wrench you are using on a lag screw from a height of, say, 8 feet, it will clobber anyone below long before you can shout, "Heads up!"

A hard hat can also prevent serious injury to yourself when you are lopping off limbs and branches. And sometimes, even if you are not working in a eucalyptus, a tree will drop a branch on its own.

Goggles

Some people object to safety goggles because they fog up, gather dust deposits, and make you sweat. But if you value your eyesight, you should consider wearing them whenever you are sawing wood with a power saw or doing anything else that has the possibility of setting loose pieces of the material you are working on.

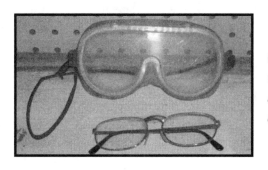

Goggles. *This pair, which can save your eyesight, feature ventilation holes in the top and sides to minimize fogging, and they are large enough to fit comfortably over a pair of eyeglasses.*

Dust Masks

You do not want to breathe in the dust created when you saw or sand any kind of wood, especially if it is arsenic-treated, pressure-treated lumber, so the proper dust mask is important.

Not all dust masks are the same. Some are meant only to give a certain amount of protection from "sweeping, cleaning the garage, and mowing," as is specified on the package. Some labels will warn, "Do not use this mask for protection against toxic or hazardous dusts such as silica, lead, asbestos, cotton, wood or grain and paint sprays, pesticides, any fumes, mists, gases and vapors." The masks that come in this package are not suitable for the construction environment, as the label makes clear. So be sure to read the label on the mask package and purchase a type of mask that will give some protection against the products of cutting and sanding wood.

Table Saws

If you have a small table saw or "contractor's saw," you will be able to make cuts more accurately and more quickly than most people can with a hand-held circular saw. The table part of the saw is also a steadier platform than that afforded by sawhorses.

The ground near a tree is seldom as level or as firm as your shop or garage floor, so be sure to use boards under the legs to level the saw and to prevent the legs from sinking into the soil. Also be sure to have another person help you when you are cutting lumber of any length, or plywood sheets.

Cutting Plywood

A ¾ inch thick, 8- by 4-foot sheet of plywood weighs anywhere from 180 to 200 pounds; with a plywood sheet measuring 4 by 8 feet, that is just too much ungainly mass for one person to handle safely and make cuts with any precision.

If you do not have another person who can help you, the next best thing is to use two or three sawhorses and use a circular saw to cut the plywood sheet to size. There are cases where you may want, say, a 2- by 2-foot piece of plywood for, perhaps, a trap door. Cut the plywood sheet on the sawhorses with a circular saw to near the dimensions you want, and finish it off with the table saw for an accurate, square shape.

Another aid to cutting sheet goods with a table saw when you do not have an extra pair of hands is to use rollers to support the leading edge of the sheet as you push the sheet across the saw. Just make sure the rollers are level, and use boards to prevent the legs from sinking into the ground.

Kickback

There is a universal law about the table saw: Sooner or later, it is going to spit a piece of wood at you. This is especially true of sheet goods such as plywood. However careful you are, someday, sometime, you are going to do something to bind the wood as it is being cut, and the saw blade is going to kick that sheet of plywood back — at speeds and force that can cause a lot of damage and injury.

Most table saws these days come with anti-kickback pawl attachments. You should use this attachment if your saw has one. For relatively small work, use a feather board to prevent smaller pieces from kicking back at you.

Refrain from standing directly in front of the blade when you are making a cut. With a little practice, you can stand at the side of the saw and direct the wood just as surely as you can when standing in front of the saw.

Push Sticks

Generally speaking, tree house construction calls for relatively large hunks of wood, so there are seldom cases when your hands are going to get very close to the saw blade. But when you are cutting trim for doors or windows, or ripping a 2x4 into a 2x2, you are well-advised to use a push stick to get the wood past the blade. The stick enables you to keep fingers well-away from the saw blade for safety concerns.

Extension Cords

Be sure to use the correct extension cord for your power tools. That little two-wire "flex" cord that may be fine for your table lamps is not fine for the amperage draw of a power saw. At a minimum, use a 16-gauge extension cord rated for outdoor use. Better for power tools is a 14-gauge cord. Remember to keep the connection ends from getting wet, which will happen if you lay them in dewy grass.

Year-Round Tree House Planning, Building, and Maintenance

As the saying goes, an ounce of preventive maintenance will prevent a ton of emergency repairs. Just as you change the anti-freeze in your car whenever fall is in the air and prune fruit trees in the winter, there are times that are best-suited for tree house maintenance tasks. If you live in the Northern Hemisphere, where the winter days are cold and short, you can:

- Sketch or photograph the tree or trees that might be possible locations for your tree house, then sketch in how the structure might be formed to work with the trees, landscape, and weather conditions.

- Clear a path to the potential tree house sites so that later access on foot or by vehicle is unobstructed.

- Cut away any branches that are dead, dying, or might be in the way of your tree house in the months between December and February.

- Take advantage of the long nights to create your plan design in sufficient detail so that you will not have to stop once construction begins because of unforeseen problems.

- Order needed construction materials and tools, and figure out where you will position safety ropes and pulleys.

- Place foundation support bolts into the dormant tree so that spring growth will secure and stabilize them; winter is actually the best time of year to do this.

In springtime, you can:

- Lay the tree house foundation and build the platform.

- Place test weights on the floor and leave in place for a few weeks, to see if there are any issues regarding leveling that need to be fixed.

- Work inside; construct the walls and roof of the tree house for later hoisting and attachment.

- Begin construction of other portions of the house, keeping them small enough to be hoisted by a pulley rope into position.

Summer, with its short-sleeve weather, is a good time to complete construction of your tree house, sit back, and watch your children have fun. It is also a good time to recoat all exposed wood with a high-quality preservative every two years. When it is time to recoat the exterior, you can use either a clear preservative, or one with a tint that will give the wood a rich, healthy look.

In the fall, your tree house should offer you some spectacular views of the leaves changing color. But there are some tasks that need to be completed in this season:

- Fill a large bucket with water and splash it over the roof to see if there are any leaks that need repairing.

- Take a close look at support beams and sliding joints to check that they are working properly without rot or serious rust.

- In October and November, drain any plumbing or pipes to avoid freezing in winter.

- Clear away any debris from the roof and platform deck.

- If any ropes or suspension bridges seem like they will restrict tree growth, remove them and replace them in another location, preferably with protective bands between the rope and bark.

Once a year, check all exposed hardware — nuts, bolts, and screws — and tighten them if they are loose. Or, if they are weakened, replace them with new hardware. Also, check the foundation for signs of deterioration or rot. Be sure that metal brackets have not moved, and if necessary, reposition or replace them.

It is imperative to notice the areas where the tree(s) come in contact with your structure, assuming there is an adequate space of two inches between the bark and the tree house. Annual tree growth should be minimal because most growth takes place at the end of the branches. But if there are any signs of direct contact between tree and tree house, you need to enlarge the hole around the tree with a jigsaw and refit the tree with a circular neoprene collar to allow tree movement and keep moisture out of the house.

There is no such thing as being too safe; a crucial part of your tree house experience is the exercise of safety and caution. It is an investment of time and energy that pays handsomely in the long run.

W.M. "Luke" Lukoskie and his daughter, Halina, who helped provide the inspiration for his tree house, visit the simple ladder steps that lead to the house.

Please see the color insert for photos of Lukosie's tree houses. (Photo by Lois Schwenesen)

Case Study: Pro Builder Collaborates with Daughters on Tree House with a Spectacular View

W. M. "Luke" Lukoskie
Halina's Triahouse
Vashon Island, Washington

W.M. ("Luke") Lukoskie, a professional builder and entrepreneur on Vashon Island, Washington, had a business slowdown in 1999, so he set his professional carpenters to work on a project that had its origins with a single 4x4 beam placed in a maple tree by his daughter Demetria.

Lukoskie made a rough sketch of how he wanted to the tree house to look and, after consulting with both his then-15-year-old daughter Demetria and then-8-year-old daughter Halina, gave it to his team so they could put it together. The name became Halina's Triahouse, in honor of both of his daughters. To avoid having to get an official building permit from King County, Lukoskie designed the structure to be 119.5 square feet, just under the threshold of 120 square feet for a permit.

The tree house is a cedar stud frame structure, insulated and capped with a metal roof. It rests in a maple tree; portions of the multi-trunk tree have been topped to provide support and to allow the structure to sit directly above the tree. All exposed beams are made of treated 4x6 lumber attached to the tree trunks with ¾-inch, stainless-steel bolts with washers, lock washers, and nuts.

The four supporting trunks are wrapped together by the lower collar tie beams, so that combined with the platform, the trunks act as a single support mechanism, Lukoskie said. The 4x8 and 4x10 support beams for the platform were bolted to the tree.

The floor (or platform) is supported by 2x6 floor joists, with a plywood sub-floor over the 2x6s. "The place is overbuilt, if anything," Lukoskie said. "The finish is immaculate; the cedar interior (is) finished like a fine cabinet."

How to promote inclusion in the tree

Notches for the beams were treated with a sealant in areas where the bark and outer cambium layer were removed. This prevents insect infestation and allows the tree to grow around the beams through the process of inclusion. It also reduces expansion pressure by tree growth that could cause the supports to fail.

Lukoskie strongly prefers big leaf maple to Douglas fir for constructing a tree house. "Big leaf maples are short and squat, versus Douglas fir trees that are tall and narrow," he said. "Fir trees have a very shallow root system; it is a wonder that they stay up at all. How many maples do you see fall in a storm? Never build a tree house in [an] alone-standing fir tree."

The location is a hillside on the west side of Vashon Island, which lies at the southern end of Puget Sound in Washington State and commands breathtaking views of both the sound and the Olympic Mountains. Although the actual elevation of the tree house from the ground is 22 feet, because of the down slope of the hill, it is as if the elevation were 70 feet, according to Lukoskie.

A hair-raising experience during construction taught Lukoskie the value of pulleys. The crew had strung a rope line between two trees, then put another line across the first to hoist things up to the work site. A complete wall had been constructed on the ground, and as it was slowly raised, the lifting rope began to fray and gradually come apart. When they finally got the wall up, it was suspended on a single strand of rope.

They were able to grab the wall and attach it, but Lukoskie still shudders to think about how much damage or even injury a fall wall could create. From that point, the crew used a standard rope and pulley for lifting objects. And there is a permanent pulley mounted on a boom outside the tree house for hoisting such items as drinks, fresh bedding, and snacks.

Clear cedar with yew accents

The interior walls of Halina's Triahouse are finished with rich cedar paneling, and cedar beams appear throughout the structure. Lukoskie used some small-paned French windows he salvaged, plus a stained glass window that he bought in London. He also used yew for window moldings.

"Yew is classified as an evergreen hardwood," he said. "To the Native Americans, the yew is considered a sacred tree. Yew is harder than oak, but very flexible. It was used to make the English longbow and by Pacific Northwest Indians to make their canoes."

Entrance to the tree house is through an 18- by 24-inch trap door, which can be locked. As an added safety feature, directly below the trap door is a plywood platform to catch falling youngsters and any other objects. The tree house includes a hi-tech electric toilet that collects and vaporizes sewage at 1,800 degrees Fahrenheit. It uses no water and no plumbing, but every so often, sludge that has accumulated on paper liners must be removed, Lukoskie said.

He noted that he was ordered to stop construction twice by King County building officials, after construction was complete. He successfully argued that the rules pertaining to tree houses were unclear.

Lukoskie said learned a couple of lessons during the 10-year process of constructing his tree house and overcoming hassles with King County:

"It is worth battling the government for something you love," he said, and "it is always worth waiting until you can afford to do what you want to do the way you want to do it, as opposed to having it right now at a lesser quality.

Lukoskie figures the materials alone for his tree house cost some $28,000, and said that he is happy to share his experiences and to help those who are building a tree house. Contact him at **http://treeandb.com**.

Building the Platform

The platform or foundation of your tree house is fundamental; every tree house has one, and yours should be both sturdy and well-fitted to the tree(s) in which it is suspended. Remember the general rule of thumb that the higher you build, the less weight you can support because of thinner trunks and branches.

To locate anchor points, where the platform will be attached to the tree, use a mason's string, a line level, and temporary blocks or supports to bring the strings level. You will use these lines to place the first few support beams in place. These are attached to the tree with a fixed or flexible anchor, and they are able to support the completed structure.

Floor joists that are positioned perpendicular to the beams will make up the frame. When the floor frame is topped with decking, it becomes the finished floor upon which you can build the walls and every other component of the tree house. You can get a good idea of the loading limits of your tree house by contacting your local building and planning office; ask for span tables and material requirements for beams or girders, floor joists, and decking materials.

Fixed Anchor

Tree movement is the major challenge to overcome in anchoring the platform. Because fixed anchors have a way of breaking under the stress of tree movement in the wind, flexible anchors are commonly used when two or more trees are involved. But for building a tree house in a single tree, a fixed anchor is simple and works well. To attach the anchor, drill a slightly oversized hole for a lag screw through a 2- by 1 ½-inch beam.

Then, drill a pilot hole into the tree that is just a bit smaller than the diameter of the screw. Put one or two thick washers on the tree side of the beam and one washer on the outside before anchoring the beam to the tree with the lag bolt. The lag screw needs to be ¾ to 1 inch in diameter, and 8 or 9 inches to reach the hardwood inside the tree.

Sliding Anchor

To make a sliding anchor, you have a choice of a slot sliding type or a bracket type design. The slot type involves putting a 3- to 6-inch horizontal slot in the beam instead of anchoring it to the tree with the lag bolt. The slot should be 1 inch wide and 3 to 6 inches long. The lag bolt should be somewhat loose to allow movement of the beam. This type of sliding anchor is appropriate for smaller-scale tree houses that are lower to the ground.

For larger, higher structures, a bracket-type anchor mounted to the tree that supports the beam from below is more appropriate. The bracket design allows the beam to slide back and forth without damage to the tree, and provides a very strong anchor for the beam. Because most hardware and lumber stores do not stock sliding anchors for building a tree house, you can either ask a welder to custom-make your anchors, or consider using a Garnier Limb, a metal anchor designed to support loading of 4,000 lbs. and manufactured by Michael Garnier, whose Out 'n' About tree house resort is featured in this book. More information on ordering and installing the Garnier Limb is available at **www.treehouses.com**, and you can learn more in Chapter 5 of this book.

Hanging Anchor

A hanging anchor provides the most flexibility of any kind of joint and is the best solution in instances where tree configuration or extreme tree movement present difficult obstacles. As a precaution, you should observe the rule that minimum branch size for hanging anchors is 6 inches in diameter.

One way of achieving a hanging anchor involves the use of two eye-through bolts to suspend the beam from a branch. Use a ⅝-inch bolt of sufficient length to extend through the supporting branch; use another eye bolt that extends through the beam, secured from below with a steel-bearing plate and nut. Then connect the beam and branch with a heavy-duty steel shackle that will permit movement in all directions.

Another possibility is a cable rig that allows the most flexibility. First, cut a 6-foot length of heavy-duty steel cable;

next, cut sufficient high-strength, rust-proof chain to encircle the beam two times, with a bit to spare. Embed a ⅝-inch lag bolt into the tree roughly 4 feet above the final elevation of the beam and perpendicular to the cable. Hang the beam in the tree at its final elevation using temporary lines.

Then bring the cable through the eye screw, with a cable thimble attached to the end. Next, secure the end with three cable clamps and wrap the chain around the beam. Securely attach the ends of the chain to the cable, using a cable thimble and three clamps. Be sure to leave adequate length of the beam to extend past the chain to ensure the beam does not slip out of the anchor. Install a strong metal bracket on the bottom of the beam to loosely catch the chain.

If your branch is 6 inches or more in diameter, drilling and placing a bolt through the branch should not harm the tree. Use one or two washers on the upper side of the branch with another washer or steel-bearing plate under the beam. You must, at all times, guard against allowing the beam, cable, or hardware to rub against the tree because this can damage the bark and allow disease to kill the tree. If you cannot provide a ½-foot of space between the beam or hanger and tree, you can put a wooden block on the tree to protect it against contact with the structure.

> **IMPORTANT:** To prevent serious injury to the tree, never hang a support cable or chain by wrapping it around a branch or the tree trunk.

Another method of making a flexible joint involves suspending the tree house by ropes secured to higher branches. Three-quarter-inch Dacron rope is sufficient to support most structures, and the rope itself provides additional flexibility. When using this method, be sure to protect the branch by placing a piece of bicycle tire or other heavy rubber under the rope where it goes over the branch. The rope is secured to the branches using bowline knots, and to the platform with ⅝-inch cast eye bolts.

Single-Tree Design

Once you have found a sturdy, straight tree trunk, you can begin to build the platform. If you are building in a single tree and the size of your platform will be, say, 10 square feet, you can start by attaching two intersecting 2x8 beams placed on top of each other. Each beam will be attached to the tree with a single fixed anchor screw, or lag bolt. To support the weight of the tree house structure, the tree should measure 5 or 6 feet in diameter at its base.

Cut the lower pair of 2x8 beams to be 120 inches, and cut the upper two 2x8 beams to be 116 inches. The four beams will form a square around the tree when installed. Use a lag screw to anchor the first lower beam to the tree; center the screw in the length of board and on the tree trunk. In a similar fashion, attach the second lower beam on the opposite side of the trunk. Be sure the two beams are level with each other, and even on the ends. Attach the upper beams on top of the lower beams in the same manner.

Next, cut two side rim joists at 120 inches and two end rim joists at 116 inches. The end joists will be parallel to the upper beams. Attach the side joists over the ends of the end joists with three 16d common nails at each point. The "d" stands for penny, as in 16-penny nail. Make sure the frame is level and square, and toenail the rim joists to both sets of already-attached beams.

The all-important knee braces that support the platform come next. Cut each 2x8 to length so it positions at a 45-degree angle from one of the inside corners of the rim joist frame to the trunk. Miter the ends of the braces at 45 degrees, then fasten the braces to the frame with adjustable framing connectors using 10d- by 1 ½-inch galvanized common nails. First, place the connectors to the frame, then attach the brace to the connector.

Install another framing connector to the tree at the bottom of each brace with two ½-inch lag screws, then attach the brace to the connector with ten 16d nails.

Next, cut two common joists to go between the side rim joists; place them halfway between the upper beams and end rim joists. Halfway between the upper beams, attach two small joists that run from the side rim joists to the tree trunk. For strength and lighter weight, use 1 ¼- by 6-inch decking boards installed perpendicular to the common joists and flush with the outside of the floor frame. Using ¼-inch gaps, fasten the decking to the common joists, rim joists, and upper beams with 2 ½-inch deck screws.

Two-Tree Design

If building between two large, mature trees 6 to 10 feet apart that demonstrate little movement in the wind, it is relatively easy to fashion an open platform supported by two 2- by 10-inch beams. These will be fixed on both sides of the two trees with fixed anchors. If built in trees at least 10 inches in diameter at a lower height, tree movement should not be an issue. Another way to build a two-tree platform between trees that do respond to high winds by swaying is to suspend it by metal cables that allow it to move.

To build a simple fixed anchor platform supported by two 2- by 10-inch beams, plan on supporting the structure at both ends with 2- by 6-inch knee braces at each end fixed to the trees. The platform decking is composed of 2- by 6-inch planks, but for making a thinner platform surface, you need to install two more common joists to reduce spacing to about 16 inches on center.

Installing the beams is another instance in which leveling is extremely important. Allow the two main 2x10 beams to extend past the trees at both ends, for later trimming. First anchor one end to the tree using a ¾-inch lag bolt with washers on both sides. Use your level to make sure the beam is perfectly level before anchoring the other end in the same way. Then, install the second beams on the other sides of the two trees. Be sure the beams are level and level with each other to avoid structural issues once you start building on the platform.

The simplest way to install the floor frame is to construct it on the ground, then lift and install on top of the beams you have just constructed. As you work on the frame, check to make sure the finished frame will fit between the two trees before you complete its construction. Cut two 2- by 8-inch rim joists at 72 inches in length, and the two side rim joists at 93 inches. Then, cut the three 2x8 common joists to be 69 inches.

Using three 16d common nails per joint, nail the rim joists over the ends of the side rim joists so the joists are flush with their top edges. Arrange the common joists evenly between the end joists; attach to the side joists with 16d nails. Place the floor frame on top of the platform beams, center, and check to make sure it is square. Next, toenail the common joists and end rim joists to the already-installed beams.

Cut the 2x6 knee braces to make sure their top ends are flush with the bottom of the rim joists, and so that their bottom ends are plumb-cut at a 45-degree angle. Each of these knee braces will start about three inches-in from the end of the rim joists, and reach down to the center of the tree trunk at a 45-degree angle. Using pairs of galvanized metal joint plates and carriage bolts, join the braces to the rim joists.

Then, cut 2- by 6-inch decking boards at a length of 96 inches to be positioned parallel with the side rim joists. After spacing these planks ¼-inch apart, fasten them to the floor joists using 3-inch deck screws. Be sure to align the edges flush with the outside of the floor frame.

Cable Supports with Fixed Anchors

To build a platform that is anchored to one tree but suspended by double-support steel cables at the other end, measure the length of the 2x6 support beams from the far side of both trees, plus 8 inches. The double-supports provide redundant safety features that should prevent the structure from collapsing under any circumstances.

The basic construction of the platform is similar to the one described above, but with extra lengths of the 2x6 beams to attach the cables to the trees.

First, drill ⅜-inch pilot holes near the ends of the long 2x6 beams. Also drill four ½-inch holes through the sides of the same beams. On the anchor side, make the holes 1 inch from the end of the platform. On the other end, measure the exact diameter of the supporting tree trunk, plus 4 inches. Mark and drill on the centerline of the 2x6 beams.

At this point, you should have on-hand four ½- by 3-inch eyebolts, four nuts, and eight washers. Place the eyebolts through the center sides with the eyes inside, leaving them loose. Then screw a temporary scrap 2x4 across one end of the 2x6 beams. Mark the center point of the 2x4 between the long beams. Screw an upright 5-foot length of 2x4 centered on the halfway mark you just made on the scrap 2x4. At 4 feet from the bottom of the marked 2x4, partially insert a long screw.

Remember that you will use two cables with the four eyebolts, so when you make the cable end loops, they must first be fitted through the eyebolt. If the cable length is off a little, there is no need to be concerned, because you can easily adjust it in the tree once it is hung.

HOW TO MAKE AN ADJUSTABLE STEEL CABLE

If you have decided you want to suspend your tree house with steel cables, as described above, here is how to make the cable yourself.

Start with 10 feet or more of ¾-inch, 840-pound test working-load cable, five matched cable clamps, and three matched thimbles sized appropriately for the cable. Do not cut the cable until later, but loop the cable around one of the thimbles, allowing the cable end to extend 4 or 5 inches to allow room for the clamps. Loosely fit one clamp on and slide it up to the thimble, then tighten the nuts. As you tighten, hold the clamp body with Vise-Grips for leverage. Attach a second clamp and the loop is complete.

To attach the cables, first remove the temporary 2- by 4-inch pieces and bolt on the two inner cables, keeping the eyes on the inside. Wait until the platform 2x6 beams straddle both tree trunks before attaching the end cables. With some help from a friend, move the platform unit between the two trees and lean one end up at about a 45-degree angle. When you have decided the height of the platform, climb up on your extension ladder and measure that height — plus 4 feet — on the platform side of each trunk. At that spot, drill a 3/8- by 3-inch pilot hole.

Unless you chose a height over 10 or 12 feet, you should be able to install a lag bolt with a washer to support the platform at both ends. Once again, ask a couple of friends to lift the platform while you are on the extension ladder installing the first lag bolt. Then repeat the procedure at the other end. Your friends may want to use two 2x4s to elevate the other end of the platform.

Once mounted, the entire platform swings side-to-side like a hammock, but when it touches the trunk on its left, the tree prevents it from moving in that direction. Your challenge is to control its movement to the right and side-to-side. A small amount of movement, say, an inch, is all right because it will get your kids' adrenaline rushing without presenting any real danger.

To restrict rightward movement, add a 45-inch length of 2- by 6-inch wood to the left end. This is, then, why you added 4 inches to the extension of the 2x6 beam frame. Allowing a gap of 1 inch, or less, between the new plank and tree trunk for growth, place two 4-inch deck screws on both sides, above and below the lag bolt pilot hole. At the other end, repeat the process, but allow a full inch space between the new 2x6 board and the tree trunk.

The new board will not control end-to-end movement of the platform; that is the job of the left side. The control for side-to-side movement is the same on both sides, and ends, of the platform structure. Because you placed the new 2x6 planks visually, and not by measuring, they may be slightly crooked. Therefore, you will need to measure each piece of sideboard individually.

Here is a design that allows adjustments for tree growth: Lay a piece of 2x6 on the platform. Put it against the new 2x6 end piece. Draw a line beneath the 2x6, where it meets the platform. Visually place the sideboard sections within ¼ inch from the tree, and put two 4-inch deck screws in both ends, as well as on the other boards of the platform.

Screw each end cable to the ends of the 2x6, then place a lag bolt with washer through the middle thimble/loop. Raise the cable up as high as you can, then give the bolt a hammer tap. Drill a 3/8- by 3-inch pilot hole. Place the lag bolt in by cranking until it is snug. Repeat the procedure for the other cables, making sure to space the lag bolts sufficiently apart so they do not induce compartmentalization in the tree.

Two Trees, Two Support Posts

Another option is to construct a square or rectangular platform using two trees and two support posts of 4 by 4 inches or 6 by 6 inches sunk in the ground, anchored in concrete, and reinforced with knee braces.

This type of construction allows for tree movement by creating a floor frame that rides on top of a support beam between the two trees. The support posts, combined with the 2- by 10-inch floor frame, make for a very solid platform; the 4- by 12-inch beam also has flexibility to move on two sliding anchors between the trees to permit tree movement.

The first step is setting the posts; begin by marking the locations for the posts on the ground, according to whatever length and width the platform will be, with a few inches of extra space to compensate for tree movement.

Make the postholes 14 inches in diameter and at least 4 inches below the frost line, if you live in an area where the ground freezes. Your local building department can inform you about posthole depth requirements. Then, fill the postholes with 4 inches of compactable gravel to allow for drainage.

Make sure you cut the posts so they will be of adequate height to support the platform at your designated height. Place the posts into the holes and use temporary bracing to bring them plumb and square to the platform layout. Fill the holes with high-quality concrete and allow to cure.

When you cut the 4x12 beam, make sure it extends at least 12 inches beyond the tree anchor points at both ends. Using a mason's string and line level, transfer level over and mark the height of the posts onto the two trees. To calculate the height of the support beam, measure 9 ¼ inches from these marks. Then install the beam using steel support brackets at both ends.

To construct the floor frame, cut the two 2- by 10-inch end rim joists at 96 inches, and cut the two side rim joists and five common joists at 117 inches. Using pairs of ½-inch lag bolts, attach the side rim joists and one end rim joist flush with the tops of the support posts. With 16d common nails, join the other end rim joist over the ends of the side joists. Then, using 16 inches on-center spacing, place the common joists and fasten them to the end rim joists using joist hangers.

Saw four 2- by 6-inch knee braces at a length of 48 inches, and then miter the ends at 45 degrees so they will fit flush to the rim joists and post corners. Using galvanized metal joining plates and carriage bolts, attach the braces to the side-end rim joists. With pairs of ½-inch lag bolts or carriage bolts, join the bottom ends of the braces to the outside of the posts.

Finally, measure the diagonal angles of the floor frame to be sure it is square. Using 2 ½- inch deck screws, install 2- by 6-inch decking perpendicular to the common joists, spaced ¼ inch apart.

Three-Tree Design

A simple tree house that offers plenty of opportunities to let your imagination take flight can be constructed in three trees. The first step is to locate three sturdy trees that are within 5 or 6 feet of each other. Building materials can be from the lumberyard, or even from found logs, provided they are long and strong enough to be attached at both ends, with at least 6 inches extra at both ends.

In either case, you will need at least 13 straight logs or round fence posts for this tree house. Bore a ½-inch diameter hole for the lag bolt 6 inches from the top of each log. Place a washer under the lag bolt and hammer the lag bolt through the post and into the tree.

If you are placing the beam in a hardwood tree, it is best to first drill a pilot hole slightly smaller than ½ inch. Then turn the screw about a quarter- or half-turn, hammer the bolt again, and use a socket wrench to turn the bolt until you have a good, strong connection. Use a level to be sure the crossbeams are all level.

Nail a short piece of wood to the third tree at a right angle to the side beams for support; cut the ends of the beams at a 45-degree angle so they will join when placed on the support. You may want to notch the side beams, log-cabin style, so they will be even and level before you add the floor beams.

Add three floor-beams, or joists, by bolting ½-inch or slightly smaller lag screws to attach them to the cross beam at the open end of the structure and to the smaller beam attached to the third tree. Then cover the floor beams with 2- by 6-inch planks, spaced ¼ inches apart; fasten them down with 16d common nails.

Repeat the process when building the roof, making sure the roof is sloped toward the rear for drainage. Next, cover the roof with asphalt or some other roofing material. Make walls using tongue-and-groove vertical planks nailed with 16d common nails into the floor and roof beams. At the small corner, you might plan to put in a trap door. For more information, go to Chapter 10.

The Garnier Limb

A technological breakthrough about a decade ago, called the Garnier Limb, has made possible construction of larger and more elaborate tree houses, such as those constructed by Pete Nelson of Tree House Workshop in Seattle.

The Garnier Limb (GL) is named for Michael Garnier, a tree house builder and pioneer whose Out 'n' About tree house resort in Takilma, Oregon, claims to have the largest collection of tree houses anywhere in the world — 14, at last count.

Garnier Limbs are heavy, steel bolts that support a load in a tree — a kind of mechanical "branch" where one is needed. The basic GL is made from machine-threaded, 1-inch heat-treated, hardened steel with a unique flange. The threaded area below the flange fits into a hole drilled with a specialized bit that allows the flange to fit flush with the heartwood.

The original flange on the GL is 3 inches in diameter and 1 inch thick. But newer designs now feature a 3-inch cylinder instead of the smaller flange, anywhere from 1 to 6 inches long. They are sometimes jokingly referred to as "beer cans," but the heavier design affords greater strength and durability.

The longer collar systems (more typically used these days) are designed to allow the tree to grow over and around that portion of the extra long collar that is left exposed on the outside of the tree. This allows better sealing and encourages the compartmentalization process that the tree biologically goes through as it recovers from the original penetration.

The Garnier Limb, a steel bolt that supports a load in a tree.

The resulting anchor point can routinely withstand loads of up to 8,000 lbs. — up to 30,000 lbs. in some heavy limbs — next to the flange, making it ideal to support heavy beams and long knee braces. Extensions can be fabricated to allow the artificial limb to be as long as is necessary. The far end of these cantilevered extensions can then be suspended or braced.

Numerous applications and variations of the basic GL system as a support and component of larger brackets that accept fitted wooden support beams have been developed and used in custom-built tree houses.

Building the Deck

In home construction, the standard deck material is 1 ¼ by 4 inches of 1 ¼- by 6-inch pressure-treated, rot-resistant lumber or cedar. The confusing 1 ¼ inch designates boards that are actually 1 inch thick, compared to the more common ¾-inch thick lumber. In building your exterior deck this way, you should plan to place the planks ⅛ to ¼ inch apart. You may want to use 16d common nails into the 2x6 floor joists to hold the board in place while you replace them with coated, square drive deck screws spaced 16 or 24 inches apart.

Another approach is to use 2- by 6-inch tongue-and-groove fir decking. Because of the thickness and structural strength of this kind of flooring, it can easily span 4 feet without support, which calls for fewer floor joists. This kind of structure is well-suited for interiors; with proper weatherized coating, it can also serve as an outside deck.

Scaffolding boards 2 by 9 inches and 13 feet in length, commonly used by painters and carpenters for scaffolding, can also be used as a decking material. Because of their heft, they are able to span longer distances than regular 2x6 lumber. These boards only need support in two places, which obviates the need to build extensive floor joists. They come with a rough surface to supply traction; a light sanding may be needed to use them for decking — without splinters.

Plywood is also an acceptable, and less costly, alternative to using decking board or tongue-in-groove lumber. An exterior grade of ¾-inch plywood is easy to install, as well, but water — even on a well-treated plywood surface — will collect on the deck in little puddles, whereas with spaced pieces of lumber, the water will drain and dissipate. In attaching the decking, use deck screws that will penetrate the floor framing by at least 1 ¼ inch. And, to allow for tree growth, leave a 2-inch gap between the decking and the tree, if possible.

Once your platform and deck are securely in place, it is time to finish the structure with vertical surfaces.

Case Study:
Are Tree Houses the Geodesic Domes of the 21st Century?

Peter Nelson
Author and tree house builder
Fall City, Washington

In the world of tree houses, there are the low-budget, high-fun tree houses mostly represented in this book — projects that can be built by adults with help from children and used primarily by children. Let us call these "starter" tree houses.

Then there are the structures built for adults by adults, complete with all the amenities one might expect to find in an earth-based home, such as electric appliances and lights, plumbing, and lavish interiors — as well as stunning exteriors.

This kind of commercially built tree house can be built in the $100,000 to $300,000 price range, according to Peter "Pete" Nelson, a professional tree house builder in Washington State and author of several books on the subject. His highest-priced tree house to date is one that sold for $700,000.

A carpenter by trade, Nelson had as many as eight craftsmen working for his business last year, before the economic downturn. Now, he has two craftsmen on staff until the economic picture brightens.

Nelson says tree houses have their greatest appeal to baby boomers who want to be closer to nature. Many use their elaborate tree houses as artist and/or writer studios, where they can escape the everyday humdrum and live in their imaginations.

The trend toward tree houses is definitely a movement, Nelson said. Almost a decade ago, Nelson was writing books, pushing and promoting the tree house concept, when the *New York Times Magazine* published six full-page photos from one of his books. That exposure did much to help launch the tree house movement, he said.

The tree house community is still relatively small, however, and there is a bit of ambivalence about publicity for the simple reason that some of these structures may conflict with local building codes — and from the perspective of the tree house owner, the less said, the better. The community of tree house designer and builders maintains a supportive attitude toward one another, even though some may be business competitors.

Nelson and his wife, Judy, ran afoul of the building and planning codes in King County, Washington, when he built two tree houses near the Raging River on 5 acres they own in an environmentally critical area.

Nelson built two tree houses as the start of a treetop bed and breakfast nature resort he hopes to complete by this summer. Nelson built both structures without any kind of building permits; bureaucracy retaliated by ordering him to tear down the structures. But widespread media coverage created a backlash against the county and convinced some politicians that tree houses are not that bad after all.

In fact, the county council enacted an ordinance permitting anyone in the county to build a tree house of 200 square feet or less within certain safety parameters defined by the planning department. And, most importantly, the county council passed an ordinance that enabled Nelson to obtain the necessary permits to continue building what he hopes will be a demonstration project to promote clean ecotourism — and to grandfather-in his existing structures.

Nelson said that upon his first contact with King County officials, he expected them to be "the enemy" but was pleasantly surprised that they wanted to work with him to make his project a reality. They became people just as excited about his project as he is, he said.

In all, Nelson has ten tree house buildings planned. Each of the structures will have to be reviewed and approved individually by the county before construction can begin.

As the tree house movement spreads around the United States and the world, people in many locations who want to build typically run into a *laissez-faire* attitude on the part of local officials, or a wall of rules and regulations that must be modified or followed exactly to build a tree house, according to Nelson.

For current information on Tree House Workshop, Nelson's project, see: **www.treehouse-workshop.com/index.html**

6

Doors, Walls, and Windows

Besides adding structural strength as well as aesthetic interest, vertical surfaces serve to make a tight enclosure that protects against all kinds of weather.

Framing Walls Built on the Ground

Although every component of your tree house will appear seamless once completed, it is a good idea to think of the project in pieces as you are planning and building. Perhaps you may have seen a prefabricated house assembled and finished on a vacant lot somewhere. Gigantic trucks arrive at the site with entire walls already built. These walls are then hoisted into place and fastened.

The same principle applies to building your backyard tree house. It is simpler to frame the walls on the ground, leaving places for windows and doors, pull them up to the platform, securely install, and then finish the walls and add windows and doors once the frame is erect.

Or, you can finish the entire wall, put into place the wall siding (plywood, for example), and then lift the whole piece into place.

Tree house walls are usually framed with 2- by 2-inch or 2- by 3-inch lumber, instead of the 2x4 used in "regular" houses to reduce weight. Remember that the shape of your roof must be figured into the outline of the frame before siding is attached and the piece is raised into place.

How high should the walls be? Although standard wall height is 8 feet, 5 to 7 feet is more kid-sized and keeps the "cuteness factor" that makes tree houses so much fun. It is also important to include windows so your tree house does not become dark and dank.

Horizontal top and bottom plates attached over the ends of vertical studs are fundamental to construction of basic walls. In a traditional rectangular four-wall tree house, two of the walls are called "through" walls, and two are designated as "butt" walls. The former overlap the latter and are longer to compensate for the thickness of the butt walls. The two through walls and two butt walls face opposite each other; this means that both walls of each type are made the same dimensions.

To prepare for installation of a window, install a horizontal sill and header between two studs for a rough opening. The rough opening for a door is defined by only a header along the top. You can make similar framed openings to encompass large tree trunks close to the house.

Once the walls and openings are finished, the next step is to consider their placement on the tree house platform.

Here Come the Studs

One of the advantages of building your walls on the ground and then raising them is that you can more accurately position the frames and studs for a clean fit.

Build your wall frames by first cutting the top and bottom plates equal to wall length, without trim and siding. Place the plates together on sawhorses, the garage floor, or the ground so that their ends are in perfect alignment. Then, using either 24-inch or 16-inch on-center spacing, mark the stud layout on the plates. Allow an extra stud on all sides of the window and door openings. These are in addition to the overall layout of studs.

Also, add an extra stud to each end of the through walls if you plan to add interior paneling or some other finish. Then you will have something to nail into once the walls are fitted together. For nailing exterior siding, more studs can be useful. Cut the studs the same length as the height of the walls, minus 3 inches for the thickness of the plates.

After placing the plates over the ends of the studs, attach them using 3-inch galvanized wood screws or with deck screws through pilot holes. You can angle the screws or nails through opposite sides of the studs and into the plates, or fasten through the plates and into the ends of the studs. Either 10d or 16d galvanized nails will work, as well as screws in this instance.

Measure up from the bottom of the bottom plate, then mark the header and sill heights on both side studs to frame a window opening. Make the rough opening 1 ½ inch wider and 2 ¼ inches taller than the finished window dimensions if you will use homemade windows to account for the window jambs of ¾-inch lumber and sills made from 2x4 lumber. Make the rough opening 1 ¾ inch wider and 2 ½ inches taller than the sash, if you will use a recycled window sash without its own frame.

After you cut the windowsill and header, install between the side studs. Check to make sure the rough opening is square. To complete the overall stud layout, install short cripple studs above the header and below the sill. Follow the same procedure when framing a rough door opening, but make it 2 ½ inches wider and 1 ¼ inch taller than the finished door opening.

Now that you have built three surfaces of the tree house, it is time to consider your roof.

Framing for the Roof

Framing the roof is necessary but fun because, by now, your tree house is really starting to take shape.

Frame the two end walls (front and back walls) to follow the roof slope if you plan to build either a gabled or shed-style roof. Houses with hip (slanted) roofs are built with four standard walls and a horizontal top plate. A cone-shaped roof requires a different kind of framing than flat walls, but it is also flat across the top.

A rise-run ratio is what carpenters and builders use to determine the slope or pitch of the roof. For example, an 8-in-12 roof rises 8 inches for every 12 inches of horizontal run. A 12-in-12 roof slopes at a 45-degree angle, for example. Knowing the roof's slope is essential to framing an end wall for a gable, shed, or roof. You can also frame your roof by using its angle to the wall instead of figuring the rise-run ratio.

Another way to do this is to define the outline of the wall by popping a chalk line onto plywood sheets on the garage floor or driveway, then measuring to the lines to determine the lengths of the pieces. For example, if the roof slope for a gable end wall is 35 degree, then the top ends of all the studs and the top ends of the two top plates must be cut at a 35-degree angle. Tweak a chalk line representing the bottom of the wall, then mark two perpendicular chalk lines to the first to outline the ends of the wall.

Because the gable end wall must be a through wall, measure up from the bottom line and mark the side lines at the total height of the wall, which will be the same as the height of the side walls. Through the middle of the wall layout, pencil in a centerline. At a 30-degree angle, cut one end of each of the two top plates and leave the other ends long.

Place the angled end plates together, meeting at the centerline and intersecting one of the wall-top marks on a sideline.

Measure up from the bottom line to determine the exact lengths of all the studs. Be sure to subtract 1 ½ inch from the stud lengths for the bottom plate, then cut the top plates so their bottom ends are flush with the outside of the sidewalls. Twang a bottom chalk line and two perpendicular sidelines to lay out an end wall for a shed-style roof. Because the end walls for this type of roof need to be through walls, mark wall heights on the sidelines. To complete the layout, mark a chalk line between those two lines. Finally, cut the top ends of the studs at whatever angle you have decided for the pitch.

After you have taken all these steps, it is on to the placement of the walls.

Preparing the Walls

After the walls are built and raised, next comes installation. One way to install the walls is to make chalk lines on the platform floor where the inside edges of the wall's bottom plates will be placed. If you are making a relatively small tree house, hoist two adjoining walls to the platform, set them on their chalk lines, then connect them through the end studs using three-inch deck screws. In the same fashion, install the other walls one at a time, then join all of the walls to the platform using 3 ½-inch screws.

If yours is a larger, heavier project, lift one wall up to the platform, align it with the chalk mark and anchor it to the platform with 3 ½-inch screws or 16d galvanized nails. Keep the wall upright with a temporary brace made of 2x4 lumber. Then raise up the second wall, join the bottom plate, and fasten both walls together with the end studs. Do the same for the last two walls. Once the walls are installed, you can use a handsaw to cut out the bottom plate where the door opening will be.

Another way is to place four sheets of ¼-inch, exterior-grade 4- by 8-foot plywood on your sawhorse and mark the cut lines for the door and windows, according to your own design and measurements. Then cut exactly 5 inches up from the bottom on both sides of the door; you will cut the remainder of the door later.

Build the frame with 2- by 3-inch lumber. Start by placing several 2x4s across your two sawhorses for support, then place an uncut piece of 2- by 3-inch framing lumber on the supports, narrow side up. Place the inside top edge of the plywood on top of the 2x3, then draw a pencil line and cut the beam to your desired length, based on the configuration of the tree house.

Using 1 ½-inch deck screws, attach the narrow side of the just-cut 2- by 3-inch beam to the plywood. Measure, mark, and cut the bottom 2x3 beam in the same manner, attaching it to the bottom of the plywood. Do not screw the plywood to the part of the 2x3 beam between the door saw cuts because that portion in the doorway will be removed once the front wall is installed on the platform.

The above description is general and intended as an example of how to construct a simple basic tree house. Builders can and should come up with their own variations that will affect measurements.

Installing the Window(s) and Doors

Although different designs and approaches to construction may vary from one tree house to another, the following is a generalized description of this phase of building that can serve as a template or tutorial.

When the top and bottom 2x3 beams are in place, turn the wall over and measure the vertical 2x3s using the butt-and-cut method, or placing a piece of wood on top of the angle to which it will be joined, and marking the cutline with a pencil. Butt-up the 2x3 beams against the top of the panel, draw a pencil line under the bottom surface of the beam, and cut to fit. Before cutting, however, measure each side separately because there will be slight variations in each. Then screw them in place on both sides of the plywood with 1 ½-foot-long deck screws.

After the four 2x3 frames are joined to the plywood front, cut out the door. Mark the top outside of the door, and save the cutout for the door. The most practical door in a tree house will open outward and be placed on one side of the front for convenience and to maximize available space. Frame the door on the inside with 1- by 3-inch strapping lumber; measure two vertical lengths using the butt-and-cut method. Attach screws from the plywood front into the strapping with short screws that will not penetrate through the plywood.

Be sure that the two new boards overlap the plywood cut lines by ¼ inch so the closed door will fit snugly into the plywood cutout and rest against the strapping boards. This will help to keep weather and damaging insects outside, where they belong. Once the verticals are in place, butt-and-cut a short strapping board across the top of the door, which should overlap the opening by ¼ inch; put the panel outside-up and cut the window shape you want. Try to keep the edges and angles precise so you can then frame the windows correctly.

Lay a 1- by 3-inch piece of lumber longer than what you need on the bottom of the cutout, which rests on your sawhorses. Clamp the plank loosely to the bottom edge of the window cutout. Check for level, then tighten the clamp. Place another piece of 1x3 perpendicular to the first piece, and mar the cut lines for both pieces of molding flush with each other and with the other vertical edge of the window. Put another piece of lumber perpendicular to the horizontal piece above.

Remove both horizontal and vertical pieces and cut to size, following your pencil outline. Of course, your frame will require four pieces — two of each for the inside and two of each for the outside. Install the bottom piece first, putting it into position using the pencil marks as guides. Then invert the panel and secure it to the plywood with two small wood screws. Return the panel to the other side and clamp the remaining pieces into place, then turn over again and attach pieces with two small wood screws.

Using Butt-and-Cut to Make Accurate Fits

The butt-and-cut method of construction has been one of the carpenter's best techniques for many years. It refers to a joint formed by placing the end of one board against another board so that the boards are at an angle (usually a right angle), forming a corner.

Now that you have completed the outside, you need to invert the panel again and repeat the process from the other side. Before attaching the inside pieces of wood to the window, cut a piece of window screen that is 2 inches wider on all sides than the hole. Staple the screen in place to position it for now. Next, attach the inside frame pieces, which will also tightly secure the screen in place. Use ¾-inch screws to reach into the outer wood but not penetrate through it.

The same process can be used to add more windows anywhere else, if you want. With the front panel lying face-up, get the piece of plywood that you cut out to make the door. First, trim a ½ inch from the bottom of the door for ease of movement. Before cutting, double-check the mark on the front top of the door to be certain you are cutting the right end. Make a frame of 1- by 3-inch strapping lumber around the outside front of the door, using short screws from the plywood side.

At this point, the outer edges of the doorframe should be flush with the edges of the plywood door panel. In the same way that you used for the front panel, measure and cut the door. Measure, then cut and attach the top and the bottom of the door with ¼-inch screws every 6 inches from the plywood side. The vertical 1x3 pieces should be made using the familiar butt-and-cut technique.

Because of inevitable slight irregularities in the cut lines for the door, you should realize that the door must be mounted in the same position in which it was originally cut from the plywood sheet, The door ought to be able to fit easily into the hole left in the plywood sheet; if it does not fit, you may need to shave the door to allow a better fit.

Before placing the door in position on the plywood front panel, make three shims (wooden spacers) about ½ inch wide by $\frac{1}{16}$ inch thick by about 2 inches. Place one shim on edge on the top edge of the door cutout, and two more on the hinge side. Next, slide the door into place, pushing it firmly up against the three shims. The gap on the other long side of the door can be up to ⅛ inch thick with no problem. If that is not the case, adjust the thickness of the three shims.

You will need to build a second frame once the door is in position and before the hinges are installed. This frame will be fashioned on the plywood front around the door opening. Lay a long 1- by 3-inch board above the door without disturbing it, keeping it ¼ inch away from the top of the door. Use two screws inserted into the 1x3 and through the plywood to hold it in place.

Then butt a new 1x3 sidepiece to either side of the new top piece. Make a pencil line along the bottom of the plywood; cut two each. Align one of the new 1x3s with the ½-inch gap from top to bottom and attach temporarily with two screws. With the other vertical piece, do the same on the other side of the plywood panel. Mark the top piece, remove the screws, cut off the ends, and screw it back into place.

Mounting the Door

Next, remove the door and three spacers, turn the front panel over, and screw the new frame into place using short screws, making certain they do not penetrate to the other side. Turn the front panel over once again, take out the temporary screws, replace the three shims, and reposition the door. The best hinges are the flat pin kind. Mount them about 6 inches from the top and bottom of the door, making sure that the pins align with the ¼-inch gap.

Ridiculous as it may seem, it is a good idea to pay careful attention to which side of the door you are attaching the hinges. Masking tape sometimes helps to get the hinge properly aligned. Once the hinge is positioned, use a center punch to make one hole and place a 1 ¼-inch deck screw halfway in. Next, punch and sink another screw partway in on the other side.

Complete the sinking of one screw, being careful that the hinge does not shift position on the last turn of the screwdriver. Repeat the process with the other screw, once again checking to be sure the tips of the screws do not penetrate all the way through. Then punch and drill the other four screws, making sure that the hinge plate has not shifted. If you should notice that one of the hinges has moved, it can be easily corrected by backing out, realigning the hinge, and driving in another screw.

It is now time to put the handle on the inside of the door; for the sake of simplicity and compatibility with the wooden tree house, you can fashion a serviceable handle out of 1- by 4-inch pine. But before you attach the handle, make sure the hinges are securely in place so the door opens and closes smoothly. If not, trimming and sanding some of the plywood should fix the problem.

With your handle sanded round — except for the side that attaches to the door frame — drill two ¼-inch pilot holes in the handle, place it on the door frame, and screw in a pair of 2-inch deck screws at about the midway point on the inside of the door frame. Without a latch, the door will be held closed by installing a screen door spring so that entry involves pushing on the door.

A spring door that will stay closed will keep out severe weather, insects, and critters that would like to make your tree house their home. You can install the spring after the house is completed and attached to the tree(s).

When you make the backside of the door, you can add a small window for light as described earlier. First cut a 4- by 5-inch piece of plywood and frame it with 2- by 3-inch strips of lumber so that the framing is flush with the edges of the plywood, but wait before attaching it to the frame until after you mount the sides of the house on the platform.

Make sure you have a cordless drill, a bunch of 1 ¼-inch deck screws, a hammer, pencil, a few extra pieces of 1x3, and tape measure on the platform before you lift the four lighter wall panels up. Place the rear wall panel in position, 4 inches-in from the back of the platform, to allow the roof overhang to be free of the tree trunk.

Build Your Own Door

Another way to provide a door to your tree house is to build it yourself. First, cut 1-inch-thick lumber to the same thickness as the wall frame, but add ¾ an inch. Cut the upper doorjamb so it spans the top of the rough door opening. Attach the jamb to the header with galvanized-finish nails, making sure the outside edge is flush with the outside of the door opening. Cut the side jambs so they fit tightly between the top jamb and the floor, then install.

Make the door fit the new frame dimensions, but leave about an ⅛-inch gap around the edges of the door. Account for any offset caused by the hinges when calculating door dimensions. One good solution for the door itself is 1- by 6-inch tongue-and-groove vertical boards. First, cut the boards to desired length and fit them together. Trim one or both sides as required to get the correct door width.

Then, cut a 1- by 6-inch Z-frame consisting of parallel planks, top and bottom connected by another plank slanted at a 45-degree angle to encompass the door as well as top and bottom bracing; cut an angled piece to connect top and bottom spans. Maximum strength will be obtained if you make the Z-brace run from the top of the door down to the lower hinge. Connect all of these pieces with screws that run through the Z-bracing into the tongue-and-groove planks.

Before installing the door, place 1-inch-thick trim along the inside of the rough door opening flush with the inside edges of the doorjambs. Use outdoor-style hinges to mount the door to the jambs; make sure the door opens and shuts without catching. Install ½-inch stops from trim material along the sides and top of the door while the door is closed. You can then add a latch or some other type of handle to secure the door.

Raising the Wall Panels

If you have another pair of helping hands to hold up the wall panel vertically, place two or three 3-inch deck screws through the 2x3 at the bottom of the wall frame and into the platform. If not, use one or more of the scrap 1x3s to serve as a brace to hold the panel upright.

Next, put one of the side panels in place and attach it to the top corner from the outside using 1 ¼-inch deck screws. At this point, it is a good idea to attach a scrap 1x3 support on each side to stop both panels from sagging or falling apart.

Carefully position yourself on the ladder on the outside of the frame and place a screw in the bottom corner of the side panel, keeping the rear backside of the panel aligned with the back of the rear wall panel. Repeat the process on the other side. Slide the front panel, with door opening, into place by first pushing or pulling the two side panels apart against the pieces of scrap lumber attached to the sides of the platform.

At the top where the front panel inserts between the two side panels, place a 2-inch deck screw to join the panels.

Screw two more screws on each side where the panel joins the platform, then place screws at 6-inch intervals verti-cally. From inside the house, place more additional 3-inch screws through the 2x3 into the platform. Be careful not to put screws into the bottom 2x3 plank that runs along the cutout for the door. You will use your handsaw to cut out that section, after all the other 2x3s have been tightly secured to the platform.

Butt-and-cut four lengths of 2x3 to complete the side panels, as well as one 2x3 to span front-to-back across the top beams. This also supports the roof. Make sure this beam is placed in the exact center of panel tops, equidis-tant from the side panels. Door-closing hardware, including the door spring and mounts, is easily available at the hardware store, and should be installed at this point. Mount this hardware inside, close to the top of the door.

In preparation for the roof, measure the distance from one side panel to the other, adding 12 inches to allow for roof overhang. Divide that number in half to find the width of each roof panel, measure from the top of the front panel to the top of the rear panel, and add 8 inches. That number will be the length of both roof panels. With these numbers, your roof will have a 6-inch overhang on the front and sides, and a 2-inch overhang in the rear.

Using the remaining plywood sheets, you can fashion a simple flat roof that will be durable. First, align one panel with one edge in the middle of the center 2x3 support beam. Place the panel so there is a 6-inch overhang in front, and two inches in the rear. Fasten this with 1 ½-inch deck screws every 6 inches. After placing the second panel, place several screws to hold it in place, then complete the screws while standing on the ladder.

A relatively simple and effective roof surface can be achieved using a brush-on fibered roof coating directly onto the plywood panels. You can also use shingles or some other type of roofing material, such as shakes.

Curved Walls

For those who want to create a tree house with something other than the basic box structure described above, curved walls and a rounded shape are a nice alternative.

From a construction and design point of view, curved walls are no different than flat walls, with top and bottom plates, studs, and openings for doors and windows.

The essential difference is that you should use a double layer of ¾-inch-thick plywood to make each of the plates — sections that form the curved wall. Plan to use 2- by 3-inch or larger studs for framing curved walls; stud spacing is determined by what kind of material you use for the exterior siding material.

To cut your plates, lay them out on the ground. Use a trammel, or a thin flat board anchored at one end with a nail or screwdriver into the ground and two holes for a pencil at the other end. Make the two pencil holes the same width as the wall studs. Distances between the pencil holes and the pivot nail determine the inner and outer radii of the curve.

Next, mark the plate outlines on sheets of ¾-inch exterior-grade plywood and cut along the line with a power jigsaw. Place the studs appropriately for the siding:

- If you are using plywood, space the studs 2 inches apart for each 12 inches of outside radius.

- If there is a 3-foot radius, you should have studs every 6 inches.

- If you are using vertical 1- by 4-inch tongue-in-groove planks, place the studs at 2 feet on center, then build 2-inch horizontal nailers for temporary support between the studs along the middle of the wall. The nailers must be cut to the same radius as the wall plates. Stagger the nailers up and down to leave room to fasten.

Curved walls will give your tree house a unique look and make it possible to create castle-like turrets, but for some, the extra time and expense may not be appropriate. Either way, your kids are sure to love the finished product.

Trim and Siding

Whether you decide to use plywood or something unusual like sheet metal for siding, the whole idea is to keep it light so it does not overstress the tree.

Exterior-grade plywood in 4- by 8-foot panels is a good choice for siding because it is readily available at the lumberyard and comes in various thicknesses from ⅜ inch to ⅝ inch. Besides being durable, plywood that is aligned vertically adds strength to the overall structure.

For tree houses, ⅜-inch plywood works well placed over studs spaced 16 or 24 inches on center. Attach plywood panels vertically so they will meet over the centers of the wall studs. Place the trim on after you have installed the plywood siding.

If you have bought the type of siding that has edges that overlap at the joints to keep out moisture, be sure to follow the manufacturer's instructions when installing. This is similar to tongue-in-groove, but in this case there is a bit of wood that is designed to slightly overlap the adjoining piece. Think of the way a lapstrake boat hull is made; this is a smaller example of the same technique. Leave a ⅛-inch gap at the joint if panel edges are square, and fill the gaps with caulking after installation is complete.

Then use galvanized nails box or siding nails to attach the siding panels to the wall framing. Place the nails at 6-inch intervals on the perimeter and at 12-inch intervals in the middle of the panels for the most secure attachment. Use a thin strip of siding to wrap the ends of through walls; end the siding flush with the ends of butt walls.

Cedar shingle siding will give your tree house a charming, rustic look that will please adults and delight children. Although cedar shingles are more expensive than some other types of siding, they are lightweight and easily installed. Tapered cedar shingles that are attached in a lap-over fashion are usually sold in 16- and 24-inch lengths and random widths. These are installed over 1- by 2-inch or 1- by 3-inch boards attached horizontally across the wall framing, sometimes referred to as spaced or "skip" sheathing.

The siding is built from bottom to top, like a shingled roof, with each row of shingles overlapping the one below. The amount of shingle left after overlapping them is known as the exposure, so spacing of the skip sheathing must be the same as the exposure. For a larger tree house, it might be a good idea to ask the local lumberyard if you should place plywood sheathing under the shingle siding to improve the strength of the siding.

When installing shingle siding:

1. Apply the skip sheathing over the wall frame with screws

2. Apply the trim

3. Install the shingle siding.

4. Space the shingles according to the shingle exposure as determined by shingle length. For example, you need a 6- or 7-inch exposure for 16-inch shingles, and an 8- to 11-inch exposure for 24-inch cedar shingles.

5. Start attaching the shingles with a double layer along the bottom of the wall after you have applied the trim.

6. Use 1¼-inch narrow crown staples with a pneumatic staple gun or 5d siding nails.

7. Allow at least 1 ¼ inch to overlap the vertical joints between shingles.

8. Continue overlapping the lower course of shingles as you work your way up to create an equal exposure on each.

9. Allow 1 or 2 inches over the exposure line of the succeeding course as you fasten the shingles.

Once the door and window jambs are installed, cut 1-inch cedar trim to finish both the tops and sides of door and window spaces. At the corners, miter or butt the boards together and use galvanized box nails to join the trim to the framing. At both ends of through walls, install vertical trim boards and wrap the trim around the end studs, flush with the inside of the wall.

Railings

Safety is why putting a sturdy railing around your platform deck is a necessity. Aesthetics and interesting design should not be top-priority, but they do add interest to the tree house. To make a relatively easy but secure railing, use 4- by 4-inch support beams that reach 36 inches above the platform floor joists.

Use ½-inch carriage bolts with washers to install the support posts at 6-foot intervals between the ends of railing runs, which are also anchored with support beams. Anchor the posts to the platform frame instead of the deck. Place support posts on both sides of staircases and other access points.

Use 2- by 4-inch or 2x6 lumber to create rails that run between the top ends of the posts horizontally. Use 3-inch deck screws to attach the rails to the inside surfaces of the upright posts. Balusters (vertical ribs) can be cut to 2 by 2 inches, long enough to overlap floor joists by 4 inches, and attached flush with the rail above. An easy way to do this is to mark the baluster 4-inch spacing on the outside of the rails. Use 2 ½-inch deck screws to join the balusters with the rails and joists.

Case Study: Timber Frame Construction Produces Elegant Structure

S. Peter Lewis
Writer/photographer/builder
Western Maine

Photographer/writer S. Peter Lewis was initially inspired to build a tree house in his native Maine years before he actually began building, when he heard TV journalist Walter Cronkite describing a lunar landing.

With the help of his father, who is an experienced carpenter and woodworker, Lewis had built a timber frame home without the use of nails or screws, held together by notched and pegged joints, and he lived in a timber frame house.

As he contemplated building his tree house, Lewis struck upon the idea of using a timber frame design for its suspension because of its simplicity, elegance, and minimal impact on the supporting tree.

What he built was a 250-square foot, two-story, timber frame tree house with spiral staircases, branch furniture, and a drawbridge. His 3-year odyssey was entertainingly and instructively described in his book, *Treehouse Chronicles: One Man's Dream of Life Aloft.*

The suspension system consists of a hexagonal steel collar suspended by steel cables that attach to supporting wooden trusses. The steel cables are joined inside a conduit bent to rest in solidly at the point where the tree trunk divides in two, at 37 feet off the ground. The collar and trusses rest 6 inches from the tree trunk, and there are no penetrations of the bark.

Wedges make collar secure

The steel cables are attached to the steel collar that encompasses the tree with welded eye-bolts. Vertical and horizontal truss timbers are connected by a section of threaded steel rod screwed into a steel pin. Wooden wedges between tree trunk and horizontal truss ends add a great deal of stability to the whole structure, according to Lewis.

Cable sections connect to the steel collar with turnbuckles. Lewis used a nested pair of rough-cut 2x4s bolted together and bolted near the bottom of each truss. This method stabilizes the lower ends of the trusses.

"The entire platform became elegant — and it had another truly wonderful benefit: I would not need to drive a single metal fastener into the trunk of the tree," Lewis said in his book. "The tree would be a passive participant, and I would never have to mar that beautifully furrowed bark with the invasion of a fluted steel drill bit."

He used 5- by 7-inch beams for the large horizontal truss members; 5- by 5-inches for vertical and diagonal truss members and rim joints on the platform (joists that formed the inner and outer ring of the hexagon); and 3- by 5-inches for the common joists. He used 4- by 4-inch white pine vertical timbers for the building, and 2- by 4-inch dimensional lumber for the rest of the framing, such as intermediate studs, rafters, and blocking.

In his book, Lewis describes the exhilaration of constructing the tree house:

"I cut and placed the first of the common joists today (May 3, 2002). These are the floor joists, which span the hexagon from one side to the other — they're set on two-foot centers. I just used hand tools. There was little noise, and I didn't frighten the birds off.

"As I worked, I pondered the wonder of working with large timbers, sharp tools, and wood thrushes; of how the joints cut so easily, and how the chisel felt solid against the heel of my hand, creating beauty with angles that met so nearly perfectly the joints all but disappeared. It felt wonderful when each joist dropped in with a 'thunk.'"

In a digital age where speed and mechanization are valued, Lewis' use of a very old-fashioned way of crafting a tree house is a refreshing reminder of the value of craftsmanship and dedication.

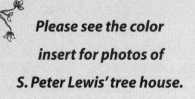

Please see the color insert for photos of S. Peter Lewis' tree house.

Raise High the Roof Beam

Whether your roof is sloped, pitched, circular, or gabled, it is an essential component of building a tree house with both structural integrity and aesthetic appeal.

Framing the Roof (Rafters)

Next to the foundation, the roof of your tree house may be the second-most important element. There are a variety of styles and methods to build your roof, and we will look at several to give you some ideas on how to get started. But first, we will review some basics.

Rafters — the lumber struts that support the roof deck or sheathing — are the primary structural component of any roof. The rafters rest atop the sidewalls and are joined at the ridge board (beam) at the roof's peak in a gable roof. Hip roof rafters also meet at a ridge beam to form a peak. Sometimes in the case of tree houses, this peak is adjacent to the tree trunk. The rafters that support a shed roof run from wall to wall, without a peak.

Though larger structures such as a house need very strong rafters spaced closely together to support the weight of the roof, smaller ones such as your tree house can be built appropriately smaller. Rafters for small tree houses can safely be fashioned with 2- by 3-inch or 2- by 4-inch beams spaces 16 or 24 inches on center. A palatial tree house that needs rafters to run more than 7 feet might require 2- by 6-inch framing, but a small tree house for kids would do well with 2- by 2-inch rafters.

Do you live in a northern location that experiences an abundance of snow in the winter? If so, you might need to speak with your local building officials to see what strengths/loads you need to consider for supporting the weight of a heavy snowfall on your tree house roof.

In making your rafters, consider making careful, precise cuts where the beam will join with the roof peak and the walls. The cut angle of the beam that meets the rook peak will be the same angle as the pitch of the rook: 35 degrees. The actual cut may be slightly different than the mathematical angle to adjust for the realities of your roof.

You will need to make a bird's-mouth cut at the other end of the rafter that will join the top plate of the wall. The name of this cut describes its appearance. The vertical cut, also called the heel cut, describes an angle with the bottom edge of the rafter that is at a 90-degree angle, less the pitch of the roof. For example, if the pitch is at a 35-degree angle, then the angle of the heel cut will be a 55-degree angle. The other cut in this rafter notch is called the seat cut because it articulates with the horizontal edge of the wall beam.

You can make a bird's-mouth cut by marking the seat-and-heel cuts with a rafter square to define the angles. A rafter square is a triangular measuring tool that can be found at any hardware store. Use a circular saw to cut one side of the rafter, then turn it over and complete the cuts with a hand saw. Be sure not to over-cut the marks, or you may compromise the strength of the rafter.

The horizontal cut, or seat cut, joins the wall at the same angle as the pitch of the roof, and is level or horizontal when you install it. Use one rafter to make test cuts and, through trial and error, find the cuts that fit best with your house. Use that as a pattern or template to mark the remaining rafters for a good fit with the wall beam. You will need to cut and fit the rafters individually if you plan to connect them to a tree.

How to Frame a Shed Roof

Working from one end wall to the other, mark the rafter layout onto the wall plates, which is the horizontal siding placed at a right angle to upright load-bearing supports. Figure the end wall top plates as rafters when you make the layout. After you have installed the regular rafters, you will make special rafters for the end-wall top plates.

To create a well-proportioned eave, cut a test rafter to length so it overhangs the sidewalls. Then set the rafter on top of the layout marks atop the two sidewalls. Where both sides of each wall intersect with the rafter, make a mark; these identify the outside ends of the seat cut in the bird's-mouth cut. To mark the bird's-mouth cuts, use a rafter square.

After making the bird's-mouth cuts, test-fit the rafter by adjusting the cut lines. As before, use the rafter as a template to mark the remaining rafters once you have achieved a good, solid fit on both walls. Then cut the rafters and use 16d galvanized common nails to attach them to the walls. The two individual outer rafters will be affixed to the top of the end wall plates. These do not have bird's-mouth cuts, but need to be trimmed to be flush with the other rafters at the top.

Building a Gable Roof

To make a ridge beam that will support a gable roof, cut a 1- by 6-inch plank to length that spans the house between the inside surfaces of the gable-end walls and gives you a structure to nail the rafter ends into. The most commonly used 2- by 3-inch and 2- by 4-inch rafters will work well with a 1x6 roof beam.

Moving from one end to the other, mark the rafter layout on the ridge beam. Include the end-wall top plates as rafters for the layout. After you install the other rafters, you will cut the special rafters. Pencil the layout on both faces of the 1x6 beam; transfer the layout onto the plate by holding the ridge beam next to each wall plate.

For an aesthetically pleasing eave, cut two pattern rafters to length so they overhang the sidewalls. Place the rafters on the sidewalls with a scrap leftover from the ridge beam; place the scrap between the top ends of the rafters. Then mark the place where both sides of both walls intersect the rafters. You will need to plane up from the wall with a straightedge because the rafter will not touch the plate on the inside of the wall. Use the marks to determine the outside ends of the bird's-mouth cut.

To mark the bird's-mouth cuts, use a rafter square; after making the first cuts, test-fit the rafters, making adjustments until the joints fit nicely on both walls. Mark the remaining rafters using the pattern rafters as a template. Install two opposing rafters, starting at one end of the house, so that the top ends of the rafters are flush with the top edge of the beam. It is all right to temporarily nail the other end of the ridge to the far wall to hold it up as you work.

Use two 16d galvanized common nails on one side of the rafter and a single nail on the other side to toenail the rafters to the endplates. Toenail through the rafters and ridge with three 10d nails at the top end. To install the remaining rafters, repeat the process.

You will install the four outer rafters on top of the end-wall plates. They need to be cut so their tops rest flush with the tops of the other rafters because there are no bird's-mouth cuts. To determine the necessary depth of the outer rafters, use a straightedge or level to plane over from the adjacent rafters and measure down to the end-wall top plates. Without the ridge beam between them, the outer rafters will meet together at the peak. Then cut the outer rafters to size and use 16d nails to join them to the top plates.

How to Frame a Hip Roof

A hip roof is a roof that slopes upward from all sides of a structure, with no vertical ends. If your objective is to build a square tree house around a central tree trunk, a basic hip roof design will work well. First, choose an angle or pitch for the roof that feels right for you. If you picked 35, for example, then cut one end of a pattern rafter at that angle. Bevel the edge of one of the outside wall corners at 35 degrees, using a circular or hand saw. You will then have a flat surface to hold the rafter.

Place the rafter on the tree and on the wall, and mark the spot on the tree where the rafter intersects. Then extend this mark around the circumference of the tree using a level. Mark the cuts for three more corner (hip) rafters by using the pattern rafter. Cut the three remaining wall corners the same way you cut the first. Placing the top ends against the tree marking and with the bottom ends resting on the walls, install the four hip rafters.

Use 3 ½-inch galvanized wood screws to attach the rafters to the tree. Use 16d galvanized common nails to fasten the hip rafters to the walls. Then hammer one nail on one side of the rafter and two toenail screws on the other side.

Delineate two mason's lines — one that runs between the top of the hip rafters just above where they intersect the walls, and the other around the ends of the hip rafter tails — to cut the interior or common rafters. As you cut the common rafters to fit, you will use these lines as a reference point. Center a full-length rafter between each pair of hip rafters to mark the common rafter layout on the top of the wall plates.

In addition, you might want to install two or more jack rafters, or short rafters that go to the walls from the hip rafters. You will need to make bird's-mouth cuts to secure the common and jack rafters to the top of the walls.

Use the Mason's lines to be sure of your measurements while you cut and test-fit the rafters, one at a time. The common rafters should be cut long, and the top ends mitered to fit the tree as much as possible. Then, cut the jack rafters and bevel the top ends so they fit flush with the side faces of the hip rafters. In this case, the objective is to get a strong joint, so do not be perturbed if your cuts are less-than-perfect. Cut the bottom end of the rafters to length so they meet the outer Mason's line, once all the joints fit together well.

As you fasten the hip rafters, also fasten the other rafters.

Making a Turret Roof

With this style roof, you will be making a pointed circular structure that needs to be firmly anchored to a 4-inch central post of appropriate length for the pitch of the roof you desire. Start with a 4-inch round post, 12 feet long. First, mark the precise center of the tower on the platform. Using the spirit level in its vertical mode, place the vertical support beam in position and hold it in place using bracers to the walls, once you have checked that it is plumb.

If the top of the support post extends above the top of the roofline, it is all right. Before attaching the plywood to the roof frame, you will cut these pieces off.

Now, you need to construct a frame from the outer edge of the tower walls to the central post, using 2- by 3-inch softwood. Mark a line around the center post at the height that you decided upon while planning the roof. At 2-feet centers, mark a line around the top ring on the plywood tower wall plate. You will be using 14 sheets of 4- by 8-foot marine grade plywood, ¾-inch thick.

Cut a length of framing material from every other one of these marks, making it 18 inches longer than the distance from your mark to the top height line around the central beam. Mark the angle you need to fit the 2x3 flush to the central beam using a combination square. Cut ten of these pieces with a miter or hand saw. By holding one of these joists against the vertical post and resting it on the outer wall, you can then mark a 1-inch groove that will enable the joist to rest in place without support. Cut a groove by hand in the same place on each of the ten pieces.

Fix these initial joists to the wall plate and to the central post equidistant with 2 ½-inch screws, remembering to skip the alternate markings you made earlier for the joists.

When you place them from one of the alternate marks not used with the first ten joists to the central post, the secondary 2- by 3-inch joists will not completely reach the post because of constricted space. Mark the angles you need to create a wedge shape between the first joists for a good fit, using the adjustable square. Repeat this process ten times; place the first joist in position resting on the top wall plate and, as before, mark for the retaining groove. You can make the length of the overhang to match the initial joists.

Using the same 2 ½-inch screws, the second round of joists can be fitted into place and secured. You will need to cover the joists with 20 sections of identical ¾-inch marine plywood. Take the measurements of one section very carefully so that just half of each of the joists is covered to allow for attachment of the next piece to the same joist.

Now, cut the marine plywood roof sections accurately; a bench saw may work better for this than a handsaw. Lift these pieces carefully to the roof one at a time, and attach them with 2-inch screws. At this point, your roof will be both conical and sturdy.

Sheathing and Shingling

Once you decide on your roofing material — whether asphalt, wood shingles, or corrugated metal — you can determine how to proceed with the sheathing, or roof deck. If yours is a small tree house, you can simply install a couple of sheets of plywood to serve as both sheathing and roof. Consider attaching 1-inch trim against the underside of the roof sheathing along the end walls to mask the outer rafters on any tree house with a shed or gable roof.

Another choice is to put down asphalt shingles over a layer of #15 building paper, on top of plywood sheathing. Although simple and durable, asphalt shingles are heavy, and their weight should be figured in to overall structural and tree strength. Use one layer of ½-inch exterior-grade plywood sheathing to make an asphalt roof. Fasten the sheathing to the rafters with 8d-galvanized nails, spaced every 6 inches on the edges and 12 inches in the field, working up from the ends of the rafters. You may overhang the sides and ends of the rafters with sheathing and shingles, if you wish.

If you want to install trim at the sides of the outer rafters, leave extra overhang there. Stagger the vertical joints between rows and cut the plywood so the vertical joints between the sheets of plywood break on the center of the rafter. Overhang the hip edges by at least 6 inches for a hip roof. Use staples to attach the roofing paper. Overlap the row below by at least 2 inches, and overlap vertical joints by at least 4 inches as you install the remaining rows. Overlap the roof peak by 6 inches on a gable roof. Then, working up from the eave, install paper on the other side.

Put a chalk line 11 ½ inches up from the eave for standard 12-inch shingles to begin shingle installation, then cut 6 inches off the end tab on the first shingle. Place the first shingle upside-down, aligning the tabs along the chalk line you just made. On gable and shed roofs, overhang the side edge by ⅜ an inch. From the bottom edge, fasten the shingle with four 2d roofing nails. Using full shingles and butting their ends together, install the rest of the starter row in the same manner.

Completely shingle one roof section at a time for a hip roof, and trim the shingles along the peak hips before you move onto the next section of shingles.

With the tabs pointing down, install the next course of shingles directly on top of the starter course. To create a 6-inch overlap of the tabs between courses, start with a full shingle. Use four nails on each shingle—one nail 1-inch in from both ends and one nail ⅝ an inch above each tab. To help you keep the shingles straight and with an even 5 inches' exposure, pop a chalk line 17 inches up from bottom of the last course you have installed.

Place the first shingle in each course by half a tab until you reach a 1 ½ inches' overhang of tabs, then begin again with a full shingle. Cut ridge-cap pieces from full shingles to shingle the peaks of gable-and-hip roofs, cutting one cap for each 5 inches of ridge, over which you need to center the caps and attach each with two nails.

Installing Cedar Shingles

Putting cedar shingles on the roof will add a nice, rustic appearance, and it is essentially the same process as installing sheathing. A word of caution: For proper drainage, do not use cedar roof shingles on roofs with less than a 3-in-12 pitch, which is approximately 14 inches.

Spacing the boards to equal the shingle exposure, sheath the roof with 1- by 4-inch skip sheathing, which are spaced boards that run the length of the roof and support the upper surface, such as cedar shingles. Check the exposure recommended by the manufacturer for the size and grade of your shingles and slope of the roof. Along the eave, install a double-starter, which is a course of shingles that overhang the roof edge by 1 to 1 ½ inches at the eave and 1 inch at the side. Starter strips, usually roll roofing or shingles, are placed along the lower edge of the roof under the bottom row of shingles to provide double coverage for the bottom row. Overlap the gaps by 1 ½ inches between courses, leaving a ¼-inch gap between adjacent shingles to allow for expansion.

Layer strips of #15 building (tar) paper into the last two courses of shingles to complete the peak of a gable roof. Use custom-beveled, 1-inch trim boards or pre-made ridge caps to cap the peak. On a hip roof, use pre-made ridge caps.

Roofing with Metal

Corrugated metal roofing is another natural for using in a tree house. Corrugated metal is available in 2-foot-wide panels of varying lengths. Get panels long enough to span each roof section so you will not have to install horizontal joints. Follow the manufacturer's instructions, as installation of metal roofing is specific to the type and maker of the material.

To install a metal roof, first build 1- by 2-inch lumber purlins, which are boards that fit between the roof rafters. Build the purlins perpendicular to the rafters. Use screws or nails fitted with self-sealing rubber washers to fasten the roofing panels to the purlins, and overlap adjacent panels at the ribs. To seal the seam, fasten through both panels. Use a sealer strip and caulk to cap roof ridges with a pre-formed ridge cap.

Waterproofing

It is important that your roof be waterproofed before attaching the walls. The best way to do this is to use liberal amounts of roofing sealant between any cracks in the sub-roofing material, such as roofing paper, and the outer layer of shingles or composition. Once you think you have got everything sealed, give it a wet run test with a garden hose. Repeat the process until it is sealed watertight before attaching walls and siding.

Adding a Skylight

A skylight serves two purposes — allowing more daylight into the tree house, and providing an excellent opportunity for stargazing at night. You can easily include a simple skylight into the design as you frame the roof. Just cut two crosspieces to fit between two roof rafters to whatever length you want, then leave a section for the skylight when you nail the sheathing to the rafters.

Allowing a 2-inch overlap on all sides, cut a section of ⅛-inch Plexiglas™ window material for the opening. Use clear silicone caulking to secure the Plexiglas to the sheathing. Slide a section of aluminum or tarpaper under the bottom edge of the Plexiglas while the caulking is still soft. To cover the roof, use roll roofing or asphalt shingles.

At the bottom of the skylight, fit the roofing beneath the aluminum flashing (which prevents rain from seeping into spaces where parts of the roof are joined at an angle), and keep applying silicone caulking at the edges where the roofing material overlaps the Plexiglas as you attach the roofing. It is important to have the roofing material overlap by at least 2 inches over the top edge of the Plexiglas.

If you want to build a skylight that opens, you need to make a box frame from 2- by 3-inch lumber so the inside of the box is flush with the inside of the hole you have made. Then, nail the box to the roof. Find a piece of ¼-inch-thick Plexiglas; cut it with a ½-inch overlap on three sides of the box. Use small bolts to attach hinges to the Plexiglas, and use flathead screws to attach the hinges to the top of the 2x3 box.

Place 2- by 2-inch flashing around the edges of the box where they touch the roof. Before you attach the final roofing material or shingles, use silicone caulking liberally and a few small nails to hold the flashing in place.

Case Study:
Sky-High and Nestled Near Mount Rainier National Park

Bill Compher
Cedar Creek Tree House
Ashford, Washington

Listening to Bill Compher tell his story about building unique tree houses that overlook the majestic Mount Rainier National Park in Washington State sounds like one of the 12 labors of Hercules.

Praised and applauded for his two tree houses and 82-foot spiral staircase leading to an observatory with incredible views of Mount Rainier and environs, Compher is pleased to give credit to his son for the 4-year staircase project.

Compher, who is also an accomplished classical guitarist and carpenter, purchased 5 acres of land in the Gifford Pinchot National Forest, which is surrounded by Mount Rainier National Park, back in the early 1980s when he worked as an accordionist, singing and playing for tourists on an antique steam train in the park.

"I used to stroll up and down the train playing the accordion," he said. "Then we would stop at the lake, and I would do a one-man concert of old railroad songs. I supported my family that way for 14 years."

During that time, he began work on the tree house that landed him, among other places, on the Oprah television show, in the pages of National Geographic, and in regional and national newspapers. He has never spent a dime on advertising, other than fees for his Web site at **www.cedarcreektreehouse.com.**

His first tree house was built from 1981-'82 in an old growth western cedar, which is 200 years old and has a 17-foot base. Compher used an umbrella foundation, with knee braces built into the tree and spread out like the arms of a bumbershoot. First, he constructed a ladder out of 2- by 4-inch planks constructed in 10-foot sections. Each section of the ladder was attached to the tree with two lag screws.

For the structural supports (knee braces), Compher cut 6-inch-thick and 15-foot-long Douglas fir poles from his property and hoisted them up and down on a pulley to the 60-foot height of the structure. He used a gasoline-powered chain saw to cut the pieces. If there was a serious misfit, he had to pulley the piece down, re-cut it, and raise it up again.

He bought dimensional, kiln-dried lumber for the joists, and used tongue-in-groove cedar to finish the inside and outside of the house. Compher used a single fir pole for the ridge beam and shiplap Douglas fir for the siding. The roof is constructed of 26-gauge galvanized metal that has weathered 27 years. He also built in four skylights, one in each bedroom, so guests can star-gaze before they go to sleep. He figures the tree house was made in equal thirds of original lumber from his property, recycled wood and fixtures, and store-bought wood.

"The cedar tree divides the loft at 60 feet up, where it is still 20 inches in diameter," he said. "The tree house was extremely difficult to build."

Compher says it took him an entire summer to complete the16- by 16-foot platform. He built the enclosed structure recessed 1 foot from the edge of the platform so he would have a place to put his feet while he worked, so the actual interior space is 15 by 15 feet.

After it was all finished and frequently used by him and his wife, the novelty began to wear off, and the structure "was abandoned for a number of years," he said. When his train musician job ended in 1995, he thought about renting out the tree house to guests, and began remodeling and renovating the structure for such use.

The house is equipped with 12-volt lights connected to a 30-watt solar panel on the roof. If sunshine is insufficient to recharge the battery, Compher must take the 65-pound battery down and replace it with a freshly charged battery using the pulleys.

He also must empty chemical portable toilets daily and carry 10 gallons of water in two 5-gallon containers each day. The water comes from a freshwater well in Ashford, which is several miles away. Compher drives to Ashford in his truck, refills the water containers, drives back, carries them some distance through the woods, and then "pulleys them up there," he said.

Compher makes ice blocks in his freezer and pulleys them up for the icebox, and lofts propane up to provide heat and cooking utilities. The house is equipped with a kitchen stove, heat, a bathroom, and four sleeping lofts. It abuts a natural mountain stream.

"It takes about four hours each day to turn [the tree house] over," he says. "It is very labor-intensive to keep it going."

Rental of the tree house is $300 per night, which Compher figures is fair because of all the maintenance work he does. Though a tree house may not be everyone's idea of a perfect vacation, for some, it represents a trip into a fantasy world. "It is not for everyone," he said. "It is a specialty, a novel experience for people who want that. Either you want to have the experience, or you do not. Most of our guests are newlyweds or married folks who are celebrating an anniversary.

"We offer a back-to-nature experience; we're not a luxury resort that's totally comfortable and cozy. It is really geared for adults. We do not allow small children, smoking, or pets."

Compher built a second tree house between1999 and 2002, as well as the Cedar Creek Observatory — access to which is provided by an 82-foot spiral staircase built around the tree built by his son, which was called a "stairway to heaven" by the National Geographic Traveler magazine. Guests are given a one-hour trek to the observatory, up the spiral staircase and across a 44-foot suspension bridge painted with colors of the rainbow.

The observatory provides views of Mount Rainier with sightings of climbers and mountain goats, as well as a vast sweep of the Nisqually Valley. Many guests take photos of the scenery during their visit to the observatory and return year after year. Although summer is "high season" for Compher's resort, visitors come throughout the year and enjoy the natural setting and opportunities for hiking.

8

Out 'n' About Treesort

Building a tree house is not always easy. The story of Michael Garnier, who is recognized as a trailblazer in building multiple tree houses, is both instructional and inspirational. Sometimes, it takes determination to keep going in the face of adversity.

Out 'n' About Treesort, Cave Junction, Oregon

When you check into Out 'n' About Treesort, you will be greeted by the Office Fairies; taken to see the Tree Fairies building the latest and greatest tree house; taken on a horseback ride by the Stable Fairies; and served a homemade breakfast by the Kitchen Fairies. This is all part of entering the real-enough fantasyland of tree houses of every imaginable size, shape, and altitude, created by owner Michael Garnier.

Out 'n' About Treesort is located in Cave Junction, a tiny town in southern Oregon. Miles of rolling green hills present a bucolic scene to the eyes alongside the obscure roads and routes that lead eastward from Grants Pass, Oregon. Everywhere is evidence of people living close to the earth in an alternative world, far from Interstate 5 where the mainstream exists. The counterculture of the '60s is alive and well in this remote region.

Garnier, a former state wrestling champion and U.S. Army medic with a whimsical sense of humor, has become a godfather to the burgeoning tree house bed-and-breakfast resort industry. He has earned his reputation by being a pioneer, both in tree house construction and by his examples of how an enterprising "treesort" operator can work around, through, and over local bureaucrats to get tree houses approved and built.

His battle with Josephine County officials in Oregon began when he built his first tree house, Peacock Perch, in 1990. In May of that year, Garnier sought a building permit from Josephine County Building and Safety to build a tree house, but county officials told him he could not even apply for the license.

During the summer of 1990, Peacock Perch was completed. In June 1992, the county planning department issued a shut-down notice and ordered Out 'n' About to cease allowing the public into the tree houses. The following October, Garnier announced that he would close because of pressure from the county. But even though he closed his doors, he was still fighting for his tree houses.

Freak Power in Rural Oregon

National news coverage and a strong show of support from the public led Garnier to reopen in 1993 and allow guests to stay in his tree houses for a donation. In June 1994, the county issued a cease-and-desist order that directed Garnier to tear down his tree house. As a challenge to county officials, on July 4, 1994, Garnier assembled 66 people, three dogs, and a cat — weighing a total of 10,664 lbs. — on Peacock Perch, demonstrating that the tree house's load-carrying capacity was far greater than required by law.

Six days later, county officials agreed to allow Garnier to let "friends" stay in his tree houses. "And we made lots and lots of friends," he said. Giddy with their success, Garnier and friends dubbed themselves "Treemusketeers" and coined a new language, substituting "tree" for any remotely similar "E" sound. Garnier applied for and received a business license to sell "noveltree" items, including T-shirts.

In fall 1994, the Oregon State Building Board agreed with Garnier that alternative stress tests could be used to prove the structural integrity of tree houses when seeking a building permit, reversing Josephine County building officials. After about a year of relative calm, Garnier was served in August 1996 with a summons from a U.S. Circuit Court to show any reason why an injunction sought by Josephine County attorneys to tear down the tree houses should not be issued,

Within a month, Garnier counter-sued the county and asked for $50,000 in damages. This generated another round of national news coverage. Surprisingly, in June 1997, the Circuit Court granted a summary judgment in favor of Garnier, who nevertheless continued to seek county approval to operate the newly created Tree House Institute of Takilma as a "vocational school." Garnier was quoted in *Smithsonian* magazine, calling local building officials "the Tree Stooges."

The Josephine County Planning Department rejected the Tree House Institute application in August 1997, and

Garnier appealed to the Josephine County Board of Commissioners. Two months later, a judge summoned county officials to explain why they had refused to process applications from Garnier.

In March 1998, the county accepted Garnier's alternative construction methods and engineering as safe and sound, and issued permits for existing tree houses, as well permits to expand the existing bed-and-breakfast permit.

Progress in Building Methods and Materials

Meanwhile, undeterred by his legal battles, Garnier continued to build tree houses and improve his methods, largely through trial and error. Although he had studied engineering and found it "boring," his background seemed just right for the challenges, even though he had no experience as a professional carpenter.

"I started as a bed and breakfast, and nobody came," he explained. "So I decided to build a tree house when I saw how much fun my kids were having playing on a tree platform I built with just boards. I just learned as I went along."

Not only did Garnier learn about methods of tree house construction — as well as how to maneuver through political and legal conflicts — he also developed a device known as the Garnier Limb (GL), a thick, heavy, steel anchor bolt that is tree-friendly and can easily support loads of 8,000 pounds. This invention has made possible the construction of larger tree houses that have all the amenities of a luxury hone.

"Our test Garnier Limb failed at 10,000 pounds of load," Garnier explained. "There is a range of strength capacity for the GL, but generally, the denser the wood, the more loads it will hold. For instance, redwood and soft pine with the GL can sustain 60 percent of the capacity of Douglas fir, and oak can sustain 150 percent of what Douglas fir can support, or about 12,000 pounds. So the overall rating for the GL is 8,000 pounds."

The Peacock Perch, once approved by county officials, needed to be reinforced with a steel truss on the back end because of damage to one of the supporting trees, caused by horses that gnawed a hole in its lower portion and "we couldn't really say that an 80-mph wind wouldn't take it down," according to Garnier.

The Mathematics of Tree Houses

Among other things, Garnier learned from this initial project the limits of wind stress and tree strength that he put into a formula: Total area of wind resistance from the tree and structure (sail) should not be increased by more than 15 percent, and the natural strength of the supporting tree should not be decreased by more than 30 percent. "The higher you go, the smaller the structure needs to be," he said.

The Swiss Family Complex tree house was built in 1992 using angled iron as a support for fixed beams; then, the Three Room Suite arose on the other side of the property next to the swimming pool. The Suite is supported by

6- by 10-inch treated Douglas fir beams attached to eight oak trees by driving ½-foot bolts into ¾-inch holes to allow flexibility without actually using a sliding platform.

From having to replace metal and wood parts due to wear and tear over the years, Garnier said he learned that tree houses are like porches in terms of their deterioration because of exposure to the weather.

In the Caval Tree, Garnier used his new device, the GL, to create a sliding platform on white oaks, which allows the trees to sway in the wind without busting the attachments to the tree house. It was built in 1999. Although he had to remove and replace the tree attachments five years later because of tree growth, he learned that metal and trees get along well.

"We found that the trees grew more under the load of the tree house, especially around the places where the Garnier Limbs are attached," he said. As the trees grew, their bark also grew over and around the limbs in a process called inclusion, which gives the attachments greater strength. Garner cites another formula: Tree sway for an 8-foot diameter tree is ½ an inch, and for a 16-foot diameter tree, it is 2 inches — an exponential increase.

Implementing the Garnier Limb

Garnier's next project was the Serendipitree, built in 1999 on three T-brackets "because we ran out of multi-trees that could be used for support. The wood-on-wood construction allows the structure to move as much as 2 inches in any direction in high winds," Garnier said.

A network of suspension bridges connects the 14 tree houses and the main lodge. They are accessible by spiral staircases and platforms that serve as rest stops and switchbacks. To prevent buckling under stress, the bridges were built with a slightly concave form. Even so, in high winds, the bridges may sway a foot or more, despite the fact they are constructed of sturdy steel cables with redundant cables and ropes. None has ever failed, Garnier said.

Garnier constructed the Treezebo in 2000, once again using his own Garnier Limb device. The GLs worked fine, although the house itself was slightly out of level, which required the placement of shims, Based on that experience, Garnier now uses a jig that helps him to find the exact center of each tree where the GLs will be placed. He also believes that even novice tree house builders can build sturdy and serviceable structures in trees of up to 12 feet in circumference.

His most ambitious tree house project is the Majestree, a double-decker structure built of Douglas fir inside and out at an elevation of 45 feet. Its lower level is a platform/deck of 130 square feet. The upper level living area is 270 square feet, with a spacious living/sleeping area, bathroom, and deluxe red oak kitchenette. Total space of both levels is 400 square feet, and corner-to-corner, it is 20 feet. Spline brackets and 4- by 6-foot knee braces support the structure.

The springtime visitor to Majestree does indeed experience a feeling of floating above the earth while gazing down into the valley below, with its horses, surrounding hills, lush green vegetation, and the sound of a rushing creek and singing birds in the near distance.

A native of Indiana, Garnier moved to Oregon after discharge from the Army in 1972 with the intention of dropping out and moving to Europe. Instead, he helped start a free medical clinic in nearby Takilma. That — plus planting trees, making furniture, and presenting "medicine shows" — sustained him until he could purchase 3.3 acres in Cave Junction in 1974. He has since added more property for a total of 9 acres. One of Garnier's inventions from those earlier days is "Dr. Birch's Yankee Doodle Picture Propeller," a wooden device with fan-shaped propeller attachments. When the subject squints closed eyes toward the sun and "Dr. Birch" spins the blades, the subject experiences lights flashing kaleidoscopically through the color spectrum. This, Garnier claims, is the forerunner of more sophisticated (and expensive) holographic re-patterning devices currently in-vogue among alternative medicine practitioners as a treatment for psychic ailments.

Because of creativity, determination, and love of the environment, Garnier has earned the respect and admiration of other tree house builders, as well as local building officials in Josephine County, Oregon.

Building Tree House Required Innovation

When Garnier decided to build a tree house for his own residence, he picked the ideal spot on his property from the standpoint of view and location. Because the site included several 300-year-old white oak trees, imagination and innovation were required to make his dream home a reality.

"To build a house the traditional way would have meant cutting through the roots, which of course could have killed the trees," he said. "Because of the age of the trees, their roots were all intertwined."

To preserve the trees, Garnier decided to anchor the 4- by 6-inch support posts below the root zone, pounding them into "refusal," or the point where they could not be pounded deeper. Then he constructed a concrete slab under the house atop the ground to serve as the garage. In so doing, he also inserted a series of tubes around the concrete pad to carry rainwater directly into the tree's roots.

"We had an engineer estimate that an 80-mph wind would generate 105,000 ft.-lbs. of torque on the structure," Garnier said. "We tested it by putting 150,000 ft.-lbs. of torque on it, and it moved just 3/8 of an inch." One of the white oaks, which thrusts through the center of the house, was sacrificed for the sake of having a sturdy support beam.

More information about Out 'n' About Treesort is available at **www.treehouses.com**.

Case Study: Author's Personal Experience at Out 'n' About Treesort

CLASSIFIED CASE STUDIES

directly from the experts

For real thrills, tree heads should make the journey to Out 'n' About Treesort in Cave Junction, Oregon. Proprietor Michael Garnier has constructed a 50-foot-high swing suspended by a steel cable that stretches between two trees on a slope. The swinger is first strapped into a harness that looks like something from the early space program, then fitted with a crash helmet. If your adrenaline is not already pumping, wait until the Tree Fairies start to hoist you upward to the full height of 50 feet. At this point, most sane individuals would cry out for help and ask to be let down again. But reckless thrillseekers such as me cannot wait to see what happens next. Upon the release of the swinger's hands, he or she goes into a complete, heart-stopping free-fall downward at a terrifying speed. Then, the thick suspension rope grabs the harness, and the rider is shot up and out like a cannonball. The swing goes out to the end of its rope — which is where the swinger feels he or she has been thrust — then shoots backward at warp speed. As your heart rate slowly slackens, the swing starts to behave like a swing rather than an Apollo 8 moon shot. Thrilling, exhilarating, death-defying, and memorable — those over 60 should think twice about this before getting themselves strapped in. Me? I'd do it again in a heartbeat.

See the color insert for photos from the Out 'n' About Treesort.

CHAPTER

The Most Important Part: Getting the Kids Involved

Involving your children in all phases of tree house planning and construction will not only give them a sense of ownership, but also unlock the key to your own childhood so you can all join together, in play and in fantasy. What better gift could a parent give his or her child than to endorse, support, and participate in their world of imagination?

That is why we recommend you engage your children at the earliest stages by asking them to cut out photographs from magazines or newspapers of houses, public buildings, or other playhouses and tree houses that they like. By doing this, you may be able to discuss what your child likes and why, and find a way to incorporate their vision into the planning and building process.

One example of how successful this can be is provided by the nautical tree house designed and built by tree house expert Dan Wright for a family in Lansdale, Pennsylvania. It is called Shore Boat because it is made of beam-and-brace construction, with railing boards curved to match the contour of the ship's hull. It was modeled after a toy ship that one of the client's children wanted to use as a design, including captain's quarters astern and a wheel amidships (see the Case Study in Chapter 10).

Such a situation, for example, provides an opportunity for a parent with sewing skills to make nautical clothes (such as pirate's clothing) to help stimulate creativity. Also, depending on the ages of your children, it is a good way to spark an interest in ships and nautical history through books, films, and Internet research. Ask your young sailors to find objects to decorate the inside of the craft according to their tastes.

If your kids are into space exploration, consider designing and building a "far out" space capsule along the lines of what Los Angeles artist/builder Ray Cirino built. His tear-shaped, personal outside dwellings (POD) are light but sturdy and easily adaptable for tree house placement (see the Case Study in the Appendix). Cirino said he feels that a tree house offers children a "nest" where they feel safe. He describes the POD as "a functional yurt with insulation." Once again, this unusual design would provide a perfect environment for kids to become space explorers and attach all kinds of knobs, dials, and switches.

Your children could recruit a friend or two from the neighborhood to join them on their explorations into deep space and to strange new worlds. Mom and Dad might assist by rigging up speakers that pipe in appropriate, computerized space exploration sounds from readily available commercial CDs.

The importance of children in tree houses is attested to by Michael Garnier, owner of Out 'n' About Treesort in Oregon — one of the largest and best-known tree house resorts in the country. Garnier said his original inspiration came from building a simple platform-type tree house for his grandson (see the Case Study in Chapter 8).

Sometimes, involvement of the kids in tree house building can have long-lasting effects. In the case of George and Sue Simon of rural Maine, their daughter Anya and son Russell helped with the construction of their first tree house when they were 8 and 11 years old, respectively. When Russell was 17, he complained that their 8- by 12-foot tree house was too small, so the parents challenged him to build his own — which he did, with the help of his father as a senior project for school (see the Case Study in Chapter 2).

Be creative. There are a number of ways you can involve your children in the tree house project. Your family is unique; ask each member of your family about what to add to the tree house to make it unique as well. A tree house is a special expression of your love and imagination, and the journey is just as important as the destination.

Setting the Stage for Play

What do younger children most enjoy with a tree house?

"Water guns, secret escapes, [and] fake walls are things kids love," said Barbara Butler, who has designed and built tree houses for an impressive list of celebrity clients — including actor Robert Redford, singer Bobby McFerrin, and actor Kevin Kline — with turreted castles and other imaginative shapes and colors. Butler likes to build her tree houses rough-hewn and uncluttered because "a tree house is more about play than habitation." See the Case Study at the end of this chapter for more information.

Some characteristics Barbara Butler has found kids really like in tree houses include:

- Play kitchens where kids can stage tea parties and gatherings with their friends, and have social interaction

- Drop-leaf tables and benches

- A sliding (or "Judas") door in the front entrance door, so kids can open it and call out, "Who goes there?"

- Zip lines that can be incorporated into any number of games, from "fireman" to "flying munchkins."

The challenge is to build a kid-friendly, safe tree house that presents the illusion of danger without actually being dangerous, Butler said.

A crow's nest or lookout point with a telescope can give the sense of being higher up to kids as they play pirate or spy games. Old-fashioned "telephones" made of empty tin cans connected by a length of twine or fishing tackle can also provide endless fun. Suggestions for games kids can play in their tree house are available at **www.random-house.com/kids/magictreehouse/activities_print.html**

In rainy or cold weather, you may also help your kids with some indoor activities they can do inside the tree house. With just a few items, such as costume jewelry, lace boots, floppy and baseball hats, and other paraphernalia, your kids can improvise for hours with various games of pretend. This enriches their communication skills and stimulates their imagination as they play the parts of superhero, fireman, TV or movie star, journalist, doctor, pirate, or warrior.

An assortment of slides, swings, above-ground tunnels, and additional playthings are also available for sale on the Internet and at playground supply retailers.

If your kids are old enough to read, a good supply of adventure and mystery storybooks, kept in a well-constructed bookshelf, will provide even more fodder for their imagination.

Specific Tree House Designs for Kids

While this book is intended for "kids of all ages," there are some designs that are tailored to younger children.

SUSPENSION TREE HOUSE

This structure is simple in design and lots of fun for active kids. The tree house rests on ¾-inch nylon or Dacron rope suspended between two trees. Get the maximum strength rope you can find; ask the hardware store for the safe working-load limits of the rope. Use 4- by 4-inch exterior-grade plywood sized appropriately for the platform; mark a 2- by 4-inch chunk in the corner of the 4x4 plywood and cut it out. For additional support and to strengthen the tree house, attach the plywood to two, 5-foot-long 2x4s. Using a keyhole saw, cut out a 16-inch square door for access, and in the center of that, cut a 4- by 1-inch hand hole. Screw two 2- by 4-inch boards across the bottom, overlapping the door by 1 inch to support the door. Drill four 1-inch corner holes through the ends of the 5-foot beams to attach the rope.

Choose two trees approximately 10 feet apart. Place the rope through one of the corner holes and tie a knot to hold it in place, then wrap the rope around one of the trees, placing it about 5 feet higher than where the platform of the tree house will eventually be suspended. Place a large nail or lag bolt into the tree, and position the rope over the nail for support. Wrap the rope several times around the tree, then tie the other end to the second hole on the side of the floor platform.

On the other side of the platform, repeat the same process. Alternate between one side and the other as you take up the slack until the platform is centered between the two trees. Once the platform is level, stabilize it by driving several small nails through the rope into the back of the tree. To make walls, cut four pieces of 2x4s into 3-foot lengths, position them vertically to the platform, and screw them into the notches you made earlier. Cut the two 1- by 4-inch top crosspieces so they overlap the vertical beams by 6 inches at each end. The other two end pieces will fit snugly against the support beams and crosspieces.

Then, add four 1- by 8-inch guard boards by screwing the pieces into the inside surface of the four beams. To frame the roof, cut three pieces of 1¼- by 4-inch panels in 5-foot lengths. Notch two of them to receive the 1¼- by 4-inch roof ridge beam, and cut the sides at the angle of the roof you want. Fit the roof ridge into the notches, and nail the roof supports to the end pieces using 10d galvanized nails. For the actual roof, cut a piece of heavy canvas to a 4- by 7-foot shape and staple to the frame using a staple gun.

TREE HOUSE A-FRAME

You can build this kid-sized 8- by 7-foot A-frame tree house by supporting it with a single tree and two 4- by 4-inch support posts, or by simply using four support posts. The basic platform and roof/walls are made by using 4- by 8-foot plywood sheets. First, measure an area with string 89.5 inches wide and 66.5 inches deep. After you have cleared the building site of rocks, brush, and any undergrowth, this area is in the center of where you will place the four support beams.

Make sure the diagonal measurements from corner to corner are the same. Where the strings intersect, dig four holes 12 inches wide and 36 inches deep to support the four 4x4, 10-foot support posts. Place the beams loosely in the holes and brace them temporarily so they are plumb.

Other materials you will need include:

- Seven 2- by 6-inch, 8-foot-long fir beams to make the floor frame
- One ¾-inch, 4- by 8-foot exterior-grade plywood for the deck
- Six pieces of 5/6-inch cedar in 8-foot lengths for the deck

- One sheet of ¾-inch exterior-grade plywood in a 4- by 8-foot shape for the floor

- Two pieces of 2- by 4-inch cedar in 10-foot lengths for knee braces

- One piece of 4- by 4-inch by 8-foot cedar for railing posts

- Four sheets of 5/8-inch exterior plywood in 4- by 8-foot dimensions for the roof and sides

- Fourteen pieces of 2- by 3-inch by 8-foot spruce for framing; one 12-foot length of 2- by 12-inch fir for the ridge board

- Two 8-foot lengths of 2- by 6-inch cedar for the railing

- Eight 8-foot lengths of 2- by 2-inch cedar for the balustrades

- Four pieces of 11/16- by 1 3/8-inch solid crown molding

- Three 2- by 6-inch cedar in 8-foot lengths for the ladder

- Two 1- by 4-inch cedar in 8-foot lengths for step supports

- Four 80-pound bags of concrete for the post beam footings

- Four 3-inch galvanized butt hinges for the windows

Using 10d galvanized nails or screws, construct the 2- by 6-inch frame 7 feet long and 8 feet wide, then hoist the platform frame over the support posts. After leveling the platform, use two ⅜- by 4-inch lag screws to attach it to the posts. Then, frame the 36-inch, 2- by 4-inch braces to the support posts. Place a large rock at the bottom of the hold for the post to rest on, and pour concrete around the base to cover the rock and bottom of the beam. Fill in the support beam holes with soil, once everything fits plumb and level. As you add soil, tamp it down with a 2x4. Fill the remaining 10 inches of the post hole with concrete and let it cure 24 hours.

To frame the structure, install three floor joists cut from the 8-foot, 2- by 6-inch fir beams; cover them with ¾-inch exterior-grade plywood. Cover the front deck with 1¼- by 6-inch deck boards, spaced at least ¼ inch apart. Construct the A-frame using 2- by 3-inch lumber that is 93 inches long, angled to the roof beam and flooring at an appropriate angle — probably a 30- to 35-degree angle — to allow sufficient height for kids.

Cut the roof panels from the 4- by 8-foot, ⅝-inch exterior plywood sheets to be 48-inches wide, allowing the bottom of the panels to extend past the floor panel by 2 inches. On one or both roof panels, you can cut a 34- by 24-inch window hole. Use 10d galvanized nails to attach the roofing panels; cut and nail the front and rear panels to the frame. Be sure to leave a 2-foot wide space for the door. Make the deck railing out of 2- by 6-inchers for the horizontal railing and 2- by 2-inchers for upright supports. Leave a 20-inch space for access to the tree house.

Make 36-inch knee braces from the 2- by 4-incher, so they will connect at a 45-degree angle with the floor joists. Using ⅜- by 6-inch galvanized lag screws, connect the top of the brace to the inside of the floor frame and the bottom to the support posts. You can make a staircase out of 2- by 6-inch beams, using 1- by 4-inch cleats to separate and support the stairs. Use ½-inch diameter wooden pegs to attach the staircase to the tree house, first drilling appropriate holes in the upper corners of the 2- by 6-inch supports and into the 2- by 6-inch floor frame.

Instead of the usual door made of heavy plywood, try making a Dutch door of four pieces of 1- by 8-inch tongue-and-groove cedar, framed on the back with 1- by 6-inch battens. Cut a 10-inch diameter hole in the top door, as well as a 13½-inch diameter circle from ½-inch exterior plywood. Cut a 10-inch diameter, 1/8-inch clear Plexiglas™ disk to go into the same size hole in the upper door. Then, cut a 13½-inch diameter circle from ½-inch exterior plywood. Screw the Plexiglas to the back of the porthole plywood and screw the porthole into the door.

Attach the two sections of the Dutch door to the doorframe using four 3-inch galvanized butt hinges. Either a standard door lock or a wood-turn catch can be used to secure the door top and bottom.

BUILD A TREE FORTRESS

Everybody knows how much kids love to build and play in forts. And what could be cooler than a tree fort?

This finished tree house will have a footprint of slightly less than 10 square feet — plenty of room for the fort and a deck with railing.

Materials for the platform include:

- Two 2- by 6-inch #2 Douglas fir in 10-foot lengths for the V-knee brace, which is a wooden brace, or support, in the shape of a V.
- Four 2- by 8-inch #2 Douglas fir in 10-foot lengths to make the floor frame
- Seven 2- by 6-inch beams of 10-foot length for the floor joists
- Two 4- by 4-inch #2 cedar in 15-foot lengths for the support posts'
- One 4- by 4-inch in 12-foot length for the knee brace and railing posts
- Two 4- by 4-foot AC plywood for the interior floor
- Nine 5/4- by 6-inch #2 cedar for the decking

Materials for the stairs:

- Two 2- by 6-inch pieces of 10-foot length for the stringers (sides)
- One 2-by 6-inch of 16-foot length for the treads (steps)

Materials for the house, or fort:

- One 4- by 4-inch #2 cedar of 10-foot length for rear corner posts
- Seven 2- by 4-inch #2 cedar of 10-foot length to make the rear and sidewall studs
- Four 2- by 4-inch #2 cedar of 8-foot length for the sill plates

- Six 2- by 4-inch #2 cedar of 8-foot length for the top plates

- Three 2- by 4-inch #2 cedar of 10-foot length for the front wall studs

- Two 2- by 6-inch #2 cedar of 8-foot length for the front header

- Four 2- by 4-inch #2 cedar of 8-foot length for cats

- Four 3/8-inch plywood in 4- by 8-foot sheets for wall sheathing

- Six 5/4- by 4-inch #2 cedar in 10-foot lengths for corners and door trim

- Two 18-inch bundles of cedar shingles for the walls

This concept is intended for construction in one tree at least 10 inches in diameter. Using a ⅝- by 5-inch lag screw, attach a 2- by 6-inch by 9 ½-foot beam, supporting it by attaching two 2- by 6-inch by 5-foot braces that form a V. From both ends of the beams, drop a plumb bob to the ground. Measure two 7-foot-long parallel lines and mark where the two support posts will be placed. Check to be sure the diagonal measurements are equal.

Dig two holes for each of the posts, approximately 1-foot wide and 3-feet deep, then put two 4- by 4-inch cedar or pressure-treated posts loosely in the holes — but do not fill the holes yet. Build the floor frame out of 2- by 6-inch joists stretched between 9-foot by 6-inch outer beams, and attached inside with metal joist hangers. Space the joists at 16-inch intervals, front to back. With the right exterior wall flush with the outer edge of the platform, there will be a 30-inch deck extension to the left of the fort which will be enclosed with a railing, as described above in "Tree House A-Frame."

Make a triangular gusset plate of ¾-inch plywood to overlap the 2- by 6-inch V-brace that attaches to the tree with a pair of lag bolts. Attach the rear section of 2- by 6-inch fir framing to the tree with a single, centered lag bolt. Construct the remainder of the floor frame out of a 2- by 6-inch fir. Sandwich two 4- by 4-inch by 5-foot knee braces between a double layer of 2- by 6-inch floor joists. Attach them to the two posts using ½- by 6-inch lag bolts.

To make a gabled roof, you will need these materials:

- Twenty 1- by 4-inch #2 pine of 8-foot length for the nailers

- Eight 2- by 6-inch #2 cedar of 12-foot length for the rafters and fascia

- One 2- by 6-inch #2 cedar of 12-foot length for the ridgepole

- One 2- by 4-inch #2 cedar in 12-foot length for the gable frame;

- One 1- by 4-inch #2 cedar of 16-foot length for trim to cover insect screen

- One 5/4- by 6-inch cedar of 14-foot length for the header trim

- Seven 24-inch bundles of hand-split cedar shakes for the roof

- One roll of 30-foot tarpaper for the roof sheathing

- One 36-inch wide roll of screen for the gables and windows

Place the 36-inch insect screen inside the roof gable consisting of a vertical 2- by 4-inch and two 1- by 4-inch trim pieces that meet the roof line at a 90-degree angle at the top, and join the bottom of the 2- by 4-inch in a V-shape. The roof ridge will extend 3 inches over the front deck to attach a pulley and basket, and the 2- by 6-inch fascia will extend 9 inches over the deck. The roof rafters of 2- by 6-inch #2 cedar should be installed approximately 14 inches on center.

Materials for the windows:

- One 2- by 3-inch by 16-foot spruce for the two front windows
- Four 2- by 3-inch by 10-foot spruce for the two side windows
- Eight pairs of screen hooks
- Eight screw eyes and hooks

For the door:

- Three 1- by 6-inch by 10-foot clear cedar
- Two heavy-duty door hinges
- One thumb latch handle
- Nails, lag screws, and joist hangers

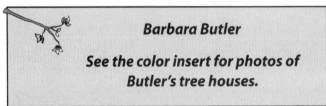

Barbara Butler

See the color insert for photos of Butler's tree houses.

Case Study: Kids' Tree House Designer Gets Kids to Experience Nature

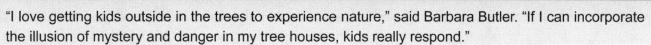

Barbara Butler
CEO/designer of Barbara Butler Artist Builder, Inc.
San Francisco
www.barbarabutler.com

"I love getting kids outside in the trees to experience nature," said Barbara Butler. "If I can incorporate the illusion of mystery and danger in my tree houses, kids really respond."

In a society where juvenile diabetes and obesity is an increasing problem, Butler wants to build tree houses that challenge kids to use their imaginations and become engaged with the natural world.

When she was a child growing up with seven brothers and sisters in upstate New York, Butler said dirt phobias and "stranger danger" were unknown fears.

The San Francisco artist/builder is famous for giving kids structures — both in trees and on the ground — where they can make their own scripts, rather than present them with limited forms and images of commercialized play. You will not find a Batman action figure in any of her tree houses.

But you will find secret passages and escapes; miniature kitchenettes; play rockets with all manner of dials and gauges; and a slightly off-center presence that Butler describes as "cartoony and buttery." At the same time, her structures have proved to be durable and safe, a place where kids can go wild with fun.

Butler's structures — which also include playhouses — resemble fine-art illustrations in older children's books of castles and other magical places. Her client list is a testament to the appeal of her—Robert Redford, Bobby McFerrin, Kevin Kline, Walt Disney Productions, Jasmine Guy, and others of similar fame.

An English major who dreamed of becoming a writer, Butler got experience in the building and construction trades alongside her brothers while she was a student. After graduate school, Barbara moved to San Francisco to let her artistic nature blossom while paying the bills with part-time construction work. While working on a play structure for singer/song writer McFerrin and his wife, Butler realized that designing and building play structures brought together her favorite things — art, building with wood, the outdoors, children, and play.

A family enterprise

Since 1996, Butler's team of 13 "play professionals" has built a total of 450 play structures, including 50 custom tree houses. She has built tree houses in Siesta Key, Florida; Santa Barbara, California; Northern California; Los Angeles, California; San Francisco, California; Texas; New York; and Minnesota.

Although some of her celebrity clients can afford more expensive and elaborate tree houses, prices range from around $3,000 to $130,000.

Money aside, Butler believes that "what really matters is the imagination of the child playing."

Some signature elements of a Barbara Butler tree house include:

• All-wood construction

• Strongly colored paint pigments

• An interior rich with gadgets, gauges, and secret escapes

• A design that she describes as "wicky-wacky," meaning eccentric and a little off-center

• Zip lines that give kids a sense of adventure

The craftsmanship that makes a Butler's structures uniquely appealing to adults and children is no accident. She strives for a soft, rounded texture on all wood surfaces to prevent splinters and to avoid wherever possible the straight-line, boxy feel of most structures. At the same time, she wants her tree houses to also be rough-hewn and not crammed with too many play objects.

"I do them pretty finished, but rough-hewn," she said. The components of a tree house are designed and built in her San Francisco workshop and assembled on-site in places such as Seoul, Korea; Las Vegas, Nevada; and other locations throughout the United States. "I do not like to make the house part too cluttered. This is more about play than habitation."

The process begins with a face-to-face visit to the client and the site, where her photographer takes lots of pictures (upon which Butler later sketches her design.) Next, Butler uses thin strips of lumber ("stick-outs") to accurately mark the outlines of the tree house in three dimensions. "We take accurate measurements and build the sections at my shop." All her tree houses are built in redwood.

That rounded, "cartoony" look

Butler achieves the smooth, rounded wood surfaces by using a 4-inch metal grinder that is useful in rounding every edge and corner; she also uses it for carving unusual shapes in wood after outlining them in chalk.

To suspend her smaller tree houses, Butler uses conventional lag bolts and knee braces, but with her larger structures she employs the heavy-duty Garnier Limb that can support loads of 8,000 lbs. or more.

Butler said she is not interested in designing mega-luxury tree houses for adults. "Adults are different," she said. "They want electrical and electronic outlets, complete bathrooms, and so on. Kids want water cannons, secret escapes, poles to slide down, and zip lines to ride." In May 2009, she was preparing portions of a tree house that she and her A-team assembled on the East Coast, while stopping at several points along the way to build more tree houses. The A-team consisted of Barbara and her husband, plus "our core group of four or five carpenters. We try to get each one built within a week. It is a kind of logistical nightmare."

One of her favorite projects involved building a tree house on the Las Vegas strip, where there were no trees. She contracted with an arborist to find a southern live oak tree in Texas, carefully removed it from the ground, packed it in a wooden box, and shipped it to Las Vegas, where it was replanted to support one corner of the Rough House Tree House. The firm also agreed to provide care and nurturing for the tree for one year. It is now open to the public and is a wheelchair-accessible facility.

10

Access and Accessories

Providing access to your tree house can also provide your kids access to their imaginations. You may want to think in terms of building a permanent sturdy staircase that kids of all ages — including adults — can use safely and add on fun ways to get in and out of the tree house, such as a rope ladder, a fireman's pole, or even circular stairs. For pure fun, you might also make a zip line, trap door, crow's nest, drawbridges, and other exciting feature for kids.

Making Double-Rung and Flat-Rung Ladders

As in most areas of design and construction, you have some choices about what kind of access ladder you want to build — double-rung or flat-rung.

In both types of ladders, build the stringers (sides) out of high-quality 2- by 6-inch beams that are free of knots. Put the end of one of the beams on the ground and the other against the tree house at whatever angle seems appropriate for climbing, possibly a 65-degree angle. Cut the bottoms of the two stringers at 65 degrees. Lay the beams against the platform at a 65-degree angle, and mark where the top ends will be cut.

Consider whether you want the stringers to be flush with the platform or perhaps extended above it and with handgrips. You can attach the stringers to the platform or to railing posts on both sides of the ladder landing once the ladder is complete. Next, cut the top ends of the stringers in the same length and shape, rounding or leaving them square, according to your preferences.

Starting at the bottom, front end of the stringers, mark the rung layout every 10 to 12 inches. The shorter rung spacing may be more appropriate for kids than the 12 inches used for adult ladders. Double-check to be sure you have uniform spacing over the entire layout, then put the stringers together and copy the rung layout to the other stringer. Use a square wooden template cut at 65 degrees to extend each layout mark across the inside face of each stringer at 65 degrees.

Assuming you will use 1 ½-inch dowels for the rungs, measure in 1¾ inch from the front and rear edges, and mark the run centers on each layout line. Then cut the rung dowels to 20 inches in length for an effective rung width of 18 inches and total ladder width of 21 inches, after the dowels are fitted into their recessed holes in the stringers. You will create these 1½-inch-around by 1-inch-deep holes by drilling at the center point of each rung with a spade bit marked with masking tape to read the depth of the hole. To be sure the rungs will fit securely, test-fit several rungs.

Before assembling and attaching the parts of your ladder, it helps to lay out all the pieces so they are nearby and ready for gluing. Coat the insides of the rung holes on one stringer with waterproof glue. Do not overdo the glue, or it will spill over and not allow the dowels to set to full depth. If necessary, use a rubber mallet to seat the dowels completely in their holes. Once all dowels are installed straight-up, glue up the other stringer and place it on the upright dowel ends. Clamp the stringers to make the ladder square, and when the glue is dry, you can attach the ladder to the platform or railing posts with screws.

Building a flat-rung ladder is similar to making a double-rung ladder, but when installed it will be a little easier to ascend and descend than a double-rung. Stair climbers should always face the ladder, whether going up or down, for safety. For another layer of safety, you can consider installing handrails on both sides of the stringers.

Follow the same procedure outlined above for making a double-rung ladder to find the ladder angle. Cut the 2- by 6-inch stringers, then measure and mark the rung layout. The layout lines for the rungs designate the top of each rung.

To fashion this ladder, cut 2- by 4-inch or 2- by 6-inch rungs to 18 inches in length. As in the case of the double-rung ladder, this size rung will produce a total width of 21 inches. Drive 3½-inch galvanized deck screws through pilot holes in the outside of each stringer into the ends of the rungs to fasten rungs and stringers together, and place the back edge of each rung so it just touches the back edge of the stringer; check to be sure the top of the rung is directly on the layout line.

Fashion 1- by 2-inch cleats, which are small blocks of wood that serve as reinforcement under the rungs. Extend the cleats to the rear edge of the stringers from just behind the front edge of the rung; drive two 2-inch screws through pilot holes on each cleat to fasten it to the stringer. Then, cut 3-inch-long blocks from a 2x2 to make the handrails, and sand until the edges are smooth. Using pairs of three ½-inch screws, attach the blocks on the front edges of the stringers at 3-foot intervals.

Use 1¼- or ½-inch dowel rods of Schedule 40 PVC tubing centered side-to-side on the blocks for the railing. Sink all screw heads below the surface of the railing for a smooth finish. To double the grip area of the railings, butt-up the railing pieces over the center of a block and join the ends using angled-finish nails.

A little extra time and care in fashioning a ladder to your tree house will pay big dividends in both safety and ease of access.

Making a Rope Ladder

Kids delight in climbing up and down rope ladders. Besides being good exercise, the ladder allows a child to indulge fantasies of boarding a pirate ship or climbing a tall mountain.

To compensate for the inherent instability of a rope ladder, one approach is to secure the bottom end of the ladder to the ground. This is done in these two simple steps:

1. Tie the two ropes together at the bottom with a locking carabiner attached, after you have hung the ladder.

2. Pour some concrete into the ground and insert an eyebolt to which you can clip the carabiner once the concrete has dried.

With this setup, it is easy to unclip the carabiner whenever you want to pull the rope ladder into the tree house.

Cut two lengths of ¾-inch nylon or manila rope, with several feet of slack, to begin your rope ladder. Cut 1- by 4-inch boards or 1½-inch diameter wood dowels at 21 inches for the rungs. Figure on one rung for every 10 inches of vertical rise in the ladder. Two inches-in from each end, centered side-to-side on the rung, drill a ¾-inch hole through each rung. Mark the ropes every 10 inches, allowing enough slack to tie off the rope ends.

Thread each rung onto the ropes, working from the top down. Then tie a simple knot under the rung so the top of the rung matches the layout lines. Continue this process until all rungs are secured to the rope.

Using a bowline knot on each rope, attach the top ends of the ladder ropes to the platform, railing, or overhead beams. For extra stability, anchor the bottom of the ladder to the ground with a concrete pier.

Wooden Stairs

Although a traditional wooden staircase may seem a bit more pedestrian than a rope ladder or a wooden ladder, it has the advantage of solid durability and — with handrails — an extra measure of safety.

To add a wooden staircase to your tree house, start by checking that the ground where the stairs will rest is firm and level. Place two 2- by 6-inch stringers against the floor frame or platform of the tree house at your desired angle. Mark the stringers where you want them cut to rest on the ground; cut off the bottom of the stringers to rest solidly on the ground. Then measure and mark the tops of the stringers at the appropriate angle, and cut to fit snugly against the side of the structure.

If you want your steps to be spaced the standard 8 inches apart, measure down that distance from the top of the stringer and mark it as point 1. Draw a level line from that point to the front edge of the stringer, and mark this as point 2. Measure from point 2 to the top of the stringer, and mark this as point 3. Then, use the measurement between point 2 and point 3 to mark on the front edge of the stringer where other steps (treads) will be located to complete the stairs.

Cut each tread precisely 16 inches in length; then, cut two 1- by 4-inch cleats at the appropriate angle to sit under the tread and against the stringer for extra support. Nail them in place to the insides of the stringers between each tread. Use 3-inch galvanized screws to screw the stringers to the treads.

Drill ½-inch holes at both top sides of the two stringers at roughly a 45-degree angle, and 1 inch from the top edge and into the floor frame. Drive the pegs into the side of the tree house so they can be removed as needed.

By building two or more riser platforms — which are small platforms that serve as resting places — you can construct a wraparound staircase that encircles the tree. Make the platforms of 2- by 6-inch timber, supported by 4- by 4-inch upright posts embedded in concrete. A hand railing atop the support beams, made of 2- by 4-inch boards, will afford safety and ease of ascending and descending.

Spiral Staircase

If you are confident of your carpentry skills, you can construct a spiral staircase. First, sink and concrete a central 8-inch post 20 inches into the ground. Then dig a circular trench 2 feet deep and within a 3-foot by 6-inch radius from the center of the post. Fill the trench with gravel. Make the treads of 2- by 10-inch lumber, cut to a length of 3 feet by 9 inches. Drill a 1-inch hole in each tread, 2 inches from its end. Drill and insert a round gate ring into the post, 8 inches from the ground.

Working upward from the central support post, position the end of the first tread above the ring and bolt it together with a 1- by 3-inch lag bolt, which is a sturdy bolt used to support heavy structures. Support the outside edge of the tread by bolting it to a post inserted into the circular trench filled with gravel. Work upward in even steps to bolt one end of each tread to the central post and the other to two upright posts, one front and the other back.

After you have bolted the last tread to the underside of the foundation, cut back all posts to a standard height. Make a hand railing out of heavy-duty rope screwed into the posts at two different heights. Once the staircase is finished, backfill the trench with concrete.

Circular Stairs

Another way to provide access is to build circular stairs that wrap around the tree, which must have at least a 20-inch-diameter trunk.

To figure out how many steps you will need, measure the distance from the ground to the top of the deck and divide by 9 inches. To determine how much lumber you will need to make a series of steps each 2 by 12 by 24 inches, multiply the number of steps by 2 feet to figure how much 2- by 12-inch lumber is needed.

Drill a single 3/8- by 4-inch screw into the tree for each step; it is attached to the underside of each step with a ½-inch-diameter bolt. Use two 34-inch-long pieces of ½-inch rebar, placed in a V-shape with its base toward the tree, inserted into a drill hole at each end for support.

One inch from the end of each step, cut a 1½-inch square hole for fastening the baluster, which will be 2 by 2 by 29 inches high. The balusters can be joined by a 1¼-inch manila rope screwed into the tops of the balusters, or by placing a railing Then drill two ½-inch-diameter circular holes on the underside of each step. Drill each hole 2½ inches from the end of the step, and you have the sockets for the rebar struts.

To install the steps, measure down 9 inches from the top of the platform to locate the top of the last step. Drill a ¼-inch pilot hole 3 inches deep into the tree for placement of a screw eye. Allow 2 inches for the thickness of your construction materials. Then, screw a 3/8- by 4 ½-inch screw eye into the tree, using a short piece of rebar to rotate the screw eye.

Use a ½- by 2 ½-inch galvanized bolt with washer and nut to attach the step to the screw eye. After inserting the two pieces of rebar into the holes on the underside of the step, mark where the bottom ends of the rebar intersects with the tree. One inch above the tree marks, drill two ½-inch-diameter holes about 1½ inch deep and fit the bottom ends of the rebar into the tree holes. You may need to drill the tree holes a bit deeper if the step is not level. Place the rebar between tree and step, and coat the holes in the bottom of the steps with industrial adhesive.

Trap Doors

If a tree house is a stepping-off point for a child's imagination, trap doors can serve as portals into a world of pirates, secret panels, and fortresses.

The typical trap door is square with its frame built into the floor joists, but here is a place where the tree house builder can also let his or her vision take wing with different shapes — rectangular, hexagonal, triangular. The main thing is to allow sufficient room for kids and (skinny adults) through the door.

Once you have decided on the size and shape of the opening, then if your floor joists are framed at 24 inches on center, simply add two side pieces between the floor joists to form a 22½-inch square hole. 16-inch joist spacing will only allow you to make a very tight 14½-inch trap door. Cut out the joist that runs through your planned opening and install two joist headers to make a larger opening. Using the same lumber you used for the frame, install one or two sidepieces between the headers to finish the opening.

On each side of the frame, add a 1- by 2-inch door stop flush with the downside of the floor, then cut the floorboards to the correct width and flush with the frame. Use 1- by 4-inch cleats on the underside of the door to fasten the boards together, or simply use a single piece of plywood. Be sure to maintain the same spacing between the trap door planks as on the floor so the trap door will blend well with the rest of the flooring.

Install hinges on the back edge of the trap door so just the hinge barrels will be above the floor level.

For Real Excitement, a Fireman's Pole

RRRrrring! The bell sounds, and firemen scramble down their pole to the waiting truck. In this case, it is kids who zoom down the fireman's pole from their tree house to the ground, ready for action.

You can incorporate a fireman's pole into your trap door by inserting the pole into a slot that you cut into the square trap door so it can be closed when not in use. Use an appropriate length of 3½-inch-diameter PVC plumbing pipe for the pole.

Allow sufficient length (approximately a foot) to secure the base of the pole into a concrete base. Cut a ¾-inch plywood collar for the pole, drill a hole for the pipe, slip it over the pipe about 18 inches from the top, and attach to rafters or the structural frame.

Plumb the pole to the ground and mark the spot, then dig a 12-inch square hole, centered where the pole will be anchored. Place the pipe into the hole and fill with concrete. Before the concrete dries, check the pole once again to be sure it is plumb.

Zip Lines: Sailing on a Wire

For sheer thrills, a zip line gives riders the exhilarating sensation of flying. Kids love them, but it is up to the adult tree house builder to ensure their safety.

The zip line is basically a high-tension steel cable suspended between a couple of trees, with a trolley that allows the rider to zip down from the tree house on a seat or handle bar. This is where the safety factor is paramount.

Some advocate strongly for a simple seat with handgrips of some kind instead of a handle bar, because of the possibility of someone losing their grip and falling face-down onto the ground. It is also possible to use a seat/handle bar combination for more improved safety.

As the rider zips down the line, he or she slows down toward the middle, partially rises on the other side, then moves backward to a slow, safe stop. There is a range of zip lines on the market from simple and hands-only, to those with seats. One type of zip line manufactured in Europe includes a braking system that uses a rubber block to retard speed.

Attachment to the suspending trees is another crucial safety concern. A typical zip line uses 3/8-inch, 19-strand stainless-steel cable that can handle several tons of weight. You affix the stainless-steel trolley before securing the line to both trees using steel ratchets and pulleys to adjust the tension on the line.

Mount the tension block with the ratchet about 8 to 10 feet higher than the platform, using an 8- by ¼-inch lag bolt. Bolt the receiving block to another sturdy tree, somewhat lower than the first bolt. Length of the line can be anywhere from 50 to 300 feet. Thread the trolley on the cable and attach the line at both ends with clamps.

Use the ratchet to wind up the cable until it clears the ground by 6 to 8 feet. Then, you can hang the seat from the trolley on a screw-fixed carabiner that allows easy removal of the seat when not in use.

Here, those of us with a large bottom may have an advantage over those without — testing the zip line. Test the joints and make sure all bolts are tight, then climb aboard and ride the zip line. If it is too low, you will know right away by your bumpy landing. Adjust the cable tension before inviting others to take a ride.

Once everyone who will use the zip line has tried it and you have made final adjustments, lock the ratchet into position and let the fun begin.

Building a Rope Bridge

Attach three vertical packers made of 2- by 4-inch lumber with 6-inch, stainless-steel screws around the back of two trees that are approximately 10 feet apart. These packers are to protect the tree from harm caused by the chain and cable. The chain is a 30-foot length of galvanized, high-tensile steel with ¼-foot links; the cable is a 30-foot length of galvanized steel cable of 3/8-inch diameter.

Use 8-inch stainless steel lag screws driven through the 2x4 and into the tree to hold the chain in place at its halfway point behind each tree, then place the two free ends of the chain on both sides of the other tree and pull tight. You can perform this operation gingerly with a pulley mounted on a four-wheel truck, or with a gang of assistants pulling a rope attached to the chain.

Use lag bolts to attach the chains to the trees, once they are as tight as possible. Cut five 1½- by 6-inch treated, ribbed redwood decking pieces in 16-foot lengths into 4-foot lengths. Drill two ½-inch holes 4 inches-in from each end of the 4-foot planks. Position the holes with a distance apart equal to three or five chain links.

Insert two 2-inch bolts through one side of the plank after positioning it above the chains. Using 1-inch washers and locking nuts, tighten these bolts below through two chain links. After widening or narrowing the second chain, put in two more bolts and tighten. Leave a gap of one link between the boards, then repeat this process to complete the entire bridge.

Protect the trunk of the first tree with the upright packers, then thread the 3/8-inch cable through all of the unused links below before wrapping it around the first tree. Bring the cable back in the other direction by threading it through the chain links on the other side of the bridge. Use three bulldog clips to join the cables together at the rear of the second tree.

Attach a 3-inch-diameter, natural-fiber rope of 24-foot length to upright posts on the deck and position it 3 feet above both sides of the bridge. Wrap a 200-foot length of ½-inch rope over and under the thicker rope and the chain and steel cable between the deck boards. Use binding string to hold the loops of the smaller rope in place.

Slides and Swings

You did not really think your kids would be endlessly amused by a static structure in the sky, did you? No, kids seem to be constantly in motion — exploring, climbing, running, and jumping. Why not build features into your tree house to satisfy your kids' need to be on-the-go?

Fortunately, there are quite a few high-quality manufactured products on the market that you can find by searching the Internet using such keywords as "children's play equipment," "kids play structures," and the like. Before you buy or build such accessories, it is a good idea to lay down a thick deposit of wood chips or other soft material in the area where the children will be jumping about like monkeys.

If possible, you might want to check the reputation of retailers who offer play structures, as well as the quality of the manufacturer. Children's safety, obviously, should always be your top priority.

You can find a variety of manufactured swings and slides, including tire swings with molded plastic tires, buoy ball swings, and toddler bucket swings, as well as slides in a plethora of sizes and shapes to fire-up a child's imagination. If you want to preserve the handmade character of your tree house, it is fairly easy to make a disk swing.

From a piece of 2- by 12-inch lumber, or two pieces of ¾-inch plywood glued together with waterproof wood glue, use a jig saw or band saw to cut a 10- or 11-inch diameter disk. Through the center of the disk, drill a ¾-inch diameter hole. Use a router and roundover bit to smooth over the edges of the disk and prevent splinters. To the tree house or tree, attach the end of a ¾-inch nylon or manila rope, using a ⅝-inch eye through-bolt.

Use a bowline knot to tie the rope to the bolt. You can also tie a few knots at different lengths above the swing to give swingers a firmer grip on the rope. This is a good time to check the clearance of the swing from the ground and any nearby objects. Tie a knot to secure the disk at a comfortable height after you thread the rope through, then trim off excess rope and prevent unraveling by binding or melting the end.

A popular variation on the traditional rectangular swing seat is the button swing, suspended by a single rope. Draw a circle of 10-inch diameter on an off-cut of 2- by 10-inch lumber. To make a button swing, use a jig saw to cut around this circular shape in a slightly eccentric amoeba form. When you have the seat cut, bevel its edges using a sharp knife. Then, drill a ½-inch hole through the center of the seat. Sand and round the edges and surfaces, give it three coats of exterior varnish, and allow it to dry. Find a very sturdy beam on the foundation and drill a ½-inch hole at least 6 inches from the bottom edge.

Push a ½-inch rope through this hole and tie a double knot on one side of the joist. When the varnish is dried, slide the other end of the rope through the button seat and tie a loose knot on the underside, so you can adjust the height of the seat according to your child's comfort level. As you can see, there are many different ways to provide access to your tree house that serve the practical purpose of entry and exit — while also stimulating the imagination.

Climbing and Hoisting

It is easy and fun to install a pulley hoist in your tree house. When you visit the hardware store, consider a pulley that will meet your needs. If your tree house plans include a pulley and bucket, then choose a pulley that will bear the weight of whatever your kids might put into that 5-gallon metal or plastic bucket. Get the corrosion-resistant pulley along with enough appropriately sized rope to reach the ground on both ends, from wherever it will be suspended.

Use a heavy lag bolt to attach the pulley to an outrigger (extended) beam that protrudes from the side of the house. You can order water cannons, periscopes, telescopes, and other accessories to complete your child's play acting and fantasies from the Internet.

Because most kids seem to have energy to spare, you probably already have seen how quickly they take to climbing — even on such objects as coffee tables and chairs. Simply placing something for your kids to climb near the tree house will serve as a sufficient invitation for them to get into motion.

The trend toward wall-climbing began several years ago when serious rock climbers erected fabricated rock wall surfaces to work out on during the cold winter months when they could not be outside climbing mountains. Today, you can purchase manufactured climbing surfaces for your kids, or make your own using sturdy plywood and bolt-on hand and footholds. Climbing nets are also available; these are nets that can be attached to upright beams from the side of the tree house.

Children also gravitate toward climbing ropes with large knots for them to hold onto. Using ¾-inch manila or synthetic rope, attach it to the tree or tree house with a lag bolt, or simply wrap it around a strong limb and tie it. Then, tie large square knots in the rope at approximately 1-foot intervals to provide children with handgrips. Once again, it is a good idea to cover the ground in the area near these devices with soft bark — for obvious reasons. The more play functions you can add to the tree house safely, the greater the utility and enjoyment of it.

CROW'S NEST

Even if your tree house does not have a nautical or pirate theme, a crow's nest is almost a necessity; it is the best perch from which to spy on an approaching enemy.

A method of adding a crow's nest to a tree house is to use half of an old wine barrel, which is readily available at most large garden-supply stores. The important factors here are to be sure the crow's nest has a level and sturdy platform, and that the ladder extends higher than the crow's nest for safe ascent and descent.

Putting a crow's nest in a tree will call for cantilevering. As is readily apparent, you must be sure to mount this type of crow's nest where the tree trunk is wide and solid enough to bear its weight and accept the lag screws.

THE COMPLETE GUIDE TO Building Your Own TREE HOUSE

This color insert section contains photos and descriptions of actual tree houses.

Insert Contents

Out 'n' About Treesort • Cave Junction, Oregon

The "Treesort," owned by Michael and Peggy Garnier, is on 36 private acres adjacent to Siskiyou National Forest. There are 25 horses, two rocking horses, four dogs, and a cat in this unusual setting.

Out'n'About is a truly unique place. Part of its uniqueness has to do with that it is a home-grown and home-based business. There are no locks on the tree house doors. It is not the Ramada or Hilton in the trees, but is a genuine, four-star "Treesort." See Chapter 8 for details on the Treesort.

TOP: Peacock Perch
Front view of Peacock Perch shows its wooden staircase and deck structure.

BOTTOM: Back View of Peacock Perch
Michael Garnier's first tree house and the subject of protracted legal. battles over an 8-year period that resulted in victory for Out 'n' About. The steel truss at the rear of the tree house was added later to satisfy county officials.

Spiral Staircase
The smooth blending of natural wood, such as the 300-year-old white oak (below) that serves as a primary support and manufactured wood to build the spiral staircase, is a highlight of the craftsmanship that went into Garnier's "Treebode," which is his own residence.

Garnier and his wife, Peggy, enjoy the deck on their three-story octagonal tree house near Out 'n' About Treesort.

Centerpiece
The powerful strength of a centuries-old white oak tree serves as the centerpiece of Garnier's "Treebode" as it rises from the spiral staircase below, up to the loft, and through the center of the roof.

Garnier's residence at Out 'n' About is a three-tiered octagonal shaped structure supported in part by centuries-old white oaks.

"Majestree"

A 60-foot-high tree house with complete living room, kitchen, and bathroom. It is, according to Garnier, the most ambitious tree house so far at Out 'n' About.

Network

A network of suspension bridges, ladders, circular stairs, and platforms connects the dwellings and main lodge.

Suspension Bridge

A close-up look at one of the steel/rope/ wood suspension bridges at Out 'n' About reveals its sturdy construction.

Caval Tree

This pre-Garnier Limb, double-decker tree house, with its fortress-like design, suggests an outpost in the old American West.

Inclusion

Trees and metal are compatible. The lag bolts at this juncture are probably redundant because of inclusion.

Solid Brace

An early corner brace built before the invention of the Garnier Limb shows some weathering but remains viable.

Out 'n' About Treesort • Cave Junction, Oregon

Platform Intersection

Access from ground level, and between tree houses, is provided by several platform intersections that also serve as anchors for a network of wooden bridges. A spiral staircase descends at lower left to another platform below.

Another Slider

A sliding tree house platform is anchored to the supporting trees, with the Garnier Limb mounted atop steel brackets that allow tree movement in the wind. Note that the two intersecting beams that form a triangular substructure are attached to two separate trees.

Another View

Long-shot view of the tree house shows how the triangular supports make possible a configuration that encompasses several trees, without injury to the trees.

Sliding Joints

Close-up view of the two intersecting triangular supports beams for Garnier's "Treebode" shows how the steel brackets rest on the Garnier Limb insertion into the tree for a sliding platform.

Michael Poris' Michigan Tree House

A network of wooden beams and steel cables suspends the tree house between three pine trees at an altitude of 26 feet.

Chains that connect the beams to the pine trees provide strength and flexibility so the tree house can move in the wind.

Supporting steel cable for this tree house is visible at deck level. The cable reinforces and suspends the staircase and landing below.

Marcel Valliere Tree House

Inverted pyramids held together with Gorilla Glue® suspend this unusual tree house designed and built by Marcel Valliere.

Dan Wright from Tree Top Builders

Twin Oaks
Two oak trees are the complete support system for this tree house built for a private client of Tree Top Builders in West Chester, Pennsylvania.

Nautical Tree House
Dan Wright of Tree Top Builders constructed this as a second tree house for the same family that commissioned another twin oak structure.

PODs and Tree House by Ray Cirino

Ray Cirino is a futuristic designer of tree houses, and this POD (personal outside dwelling) shows how the hexagonal structure can be used. It can also be adapted for use as a tree house.

Hollywood Tree House
The unique method of suspension that designer Cirino uses on his tree houses is evident in the construction of this tree house in Hollywood, California. Rebar supports suspend the structure and rest on a huge rubber gasket that encircles the palm tree.

Tree House Bunk Bed
Graceful, sweeping, curved lines enhance this child's bunk bed and provide an extra measure of safety in this design by Cirino.

Tree House Interior
Cirino carries his curved lines into the interior of his tree houses, avoiding the straight and squared line wherever possible.

Post Ranch Inn Tree House • Big Sur, California

The tree houses at Post Ranch Inn thrust the visitor directly into nature while still providing luxury. This picture shows the post suspension system and the location of the structure within the branches of a California live oak.

The Post Ranch Inn makes efficient use of a small space with a visual sense of openness and elegant redwood paneling.

Some of the tree houses at Post Ranch Inn are equipped with elegant spas and beautiful views of the forest.

George Simons Tree Houses

The first tree house built by Simons and Woros (left) is 8- by 12-feet and 15 feet above ground; the second house (right) is 12- by 12-feet and is 25 feet from the ground.

Sue Woros and George Simons enjoy the rustic setting of their first tree house.

The second tree house the Simons built is complete with a spiral staircase and wrap-around porch.

17-year-old Russell Simons designed and built the second tree house with the help of his father, George. Russell received high school credit for his project and recently completed installation of the wood paneling on the upper floor and carpeting on the bottom floor.

George Simons Tree Houses

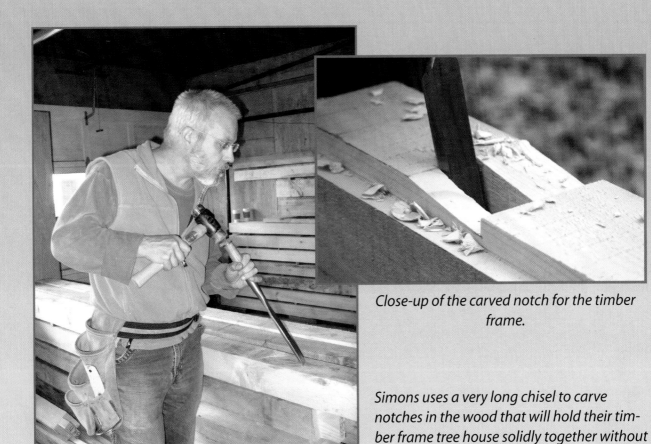

Close-up of the carved notch for the timber frame.

Simons uses a very long chisel to carve notches in the wood that will hold their timber frame tree house solidly together without the use of any hardware.

Master bedroom for the second tree house built by Simons and Woros is 12- by 12-feet with a stained glass window, small sitting area, wood burning stove, clerestory windows, spiral stairs and bridge. It is suspended from a maple and beech at about 25 feet above the ground.

George Simons Tree Houses

Simons and Woros used timber frame construction to build the frame of their third tree house, hexagon-shaped with "wings," a tower in the center, and built to accommodate six to eight adults. When completed, the structure will have a kitchen and seating area large enough to handle a large group — indoors or outdoors — queen-sized bed, and trapdoor.

This is the joinery in the center of the hexagon's ceiling that serves as the floor of the tower. This timber frame house has 340 hand-carved joints, with 53 lap joints, 103 shouldered mortise-and-tenon joints, 60-birdmouth joints, and three shoulder dovetail joints.

John Farless Tree House

The Farless family (from left: Jennifer, Brenn, and John) enjoys construction on their backyard tree house

Halina's Triahouse by W. M. Luke Lukoskie, Named After His Daughters

The view from Lukoskie's tree house is a breathtaking picture of Puget Sound.
(Photo by W. M. Luke Lukoskie.)

Luke Lukoskie of Vashon Island, Washington, built a tree house for his two daughters, Halina and Demetria, in a stand of maple trees overlooking Puget Sound on the eastern side of the island.

The high-intensity heat toilet in Lukoskie's tree house essentially vaporizes waste so it only needs to be collected every few days.

The living/dining area is compact and wood-paneled, and includes a stained glass window (upper right) that Lukoskie salvaged from a building demolition.

Lukoskie's design includes floor-to-ceiling wood paneling, plus a cozy sleeping loft.

S. Peter Lewis Tree House

Tree house under construction shows evidence of wind damage in the torn plastic wrap around door and window openings.
(Photo by S. Peter Lewis)

Peter Lewis cuts a mortise in the large platform foundation for his tree house on an early May morning. The platform was constructed on the ground, then lifted up into the tree.

Brad Mullins (right) and John Filmore-Patrick work intensively on attaching the siding. (Photo by S. Peter Lewis)

Barbara Butler Tree Houses

Bel Air Bungalow
This tree house, designed by Barbara Butler, shows a creative, artistic tree house
for children. This can serve as inspiration when building a tree house. Do not
limit your imagination. It is possible to create a whole new world just by painting
animals on a tree house.

Majestic Tree House
This tree house by Butler
is a great example of how
different color paints
can be used to appeal to
children. Also, see how
adding a rope climb,
fireman's pole, swing,
and slide can make a tree
house more entertaining.

Barbara Butler Tree Houses

Casa de Arbol
Butler designed this tree house to surround an old sycamore tree, whose thick, gnarled branches impart a sense of adventure for children.

Drop Leaf Table
A small drop leaf table with drop leaf chairs nestles next to a child-sized window in this imaginative use of space in a Butler tree house.

Sycamore Treefort
Tree house and sycamore tree become one in this Butler-designed structure.

Case Study:
His Business is Making Kids' Dreams
Come True

Dan Wright
Tree Top Builders
West Chester, Pennsylvania

Dan Wright of Tree Top Builders in West Chester, Pennsylvania, likes to say he is in the business of making kids' dreams a reality.

"It is wonderful to send a child down a zip line for the first time and hear them scream with delight. A tree house to children represents joy and freedom," he said.

In fact, a substantial portion of his business as a builder of tree houses is specifically geared for children. The excitement is contagious whenever he builds a tree house with his partner, Gary Koontz, Wright says.

"Usually when I am working on a tree house in someone's backyard, when the kids get home from school, they run straight from the school bus to see what we have done," said Wright. "To them, it is their dream. They have to go to school, do their homework, and do their chores. But a tree house is something they can call their own place."

Wright built his own tree house with his father when he was 8 years old. Then, as a teenager, he built another, with a 41-foot-high platform and a 250-foot zip line. After college, he framed and trimmed as a carpenter for a while as he attempted to get his tree house business started.

One of his earliest commercial tree house projects was building a hexagonal tree house on ash tree leaders, he recalls. From the beginning, Wright placed a high premium on coming up with an individual design that suits each customer's needs — as well as an emphasis on safety.

An autodidact, Wright says he learned a great deal about tree houses from research on the Internet and from available books. By 2004, he launched Tree Top Builders, with most of his clients coming from a triangular area, whose points would be Maine, North Carolina, and Wisconsin.

He has also built tree houses in California, Connecticut, Texas, Tennessee, West Virginia, Illinois, Virginia, North Carolina, New Hampshire, Maryland, Delaware, New Jersey, Rhode Island, Indiana, and Florida.

Travel to other states can add as much as $25,000 to construction costs, Wright said. Most of the tree houses he has built for kids in the Philadelphia area range in cost from $5,000 to $10,000. Larger adult tree houses are in the $15,000 to $40,000 price range, but he has also constructed tree houses for wealthy and famous clients in the "multi-hundreds of thousands of dollars" price range.

Like many other builders, Wright has been affected by the economic downturn but is beginning to see a slow recovery. Depending on demand, Wright usually employs one to four workers.

So does a well-constructed tree house have any impact on the retail real estate value of a home? "A tree house is like a swimming pool," Wright says. "Some like them and will pay more for them; some don't and won't. If a tree house is built to good safety standards with a good design, it can add a lot of value."

Wright confesses that, at the direction of clients, they often do not get permits after the pros and cons of doing so are explained. Many of his tree houses are suspended by high-strength aircraft cables with an artificial limb systems — a practice that he admits is controversial.

Wright says he feels blessed to have Koontz as his machinist because of his creativity and skills both in metal and woodworking. Koontz produces all the steel artificial limbs, brackets, and other hardware, and brings "a real creative flow to woodworking," Wright said. "He always comes up with something that we didn't plan. It's a joy to see what he devises."

 See the color insert for photos of Dan Wright's tree houses.

Tree House Plans

Constructing a tree house from the ground-up might initially seem like a daunting task, but with the proper plans, it is simply a matter of following step-by-step directions. This chapter will depict every tool, measurement, and instruction necessary to guide you through the creation of 15 different varieties of tree house structures, from a simple tree house to one complete with dueling slides and a swinging bridge. The plans included are:

- Simple Deck Tree House
- Kiddie Cabin Tree House
- Bird Watcher's Paradise Tree House
- Double-Deck Cabin Tree House
- Two-Story Tree House
- Tree House Landings
- King & Queen Tree House
- Dueling Slides & Swinging Bridge Tree House

- Western Fort Tree House
- Pirate Ship Tree House
- Cabin in the Sky With Wrap-Around Deck House
- A-Frame Cabin Tree House With Loft
- Rocket Tree House
- Funhouse Tree House
- Teepee Tree House

Straight from the custom designs of Randy LaTour, an autocad mechanical design expert from Peachtree City, Georgia, the following illustrated tree house plans all feature creative, original tree house construction ideas to instruct — and inspire.

Simple Deck Tree House Plan

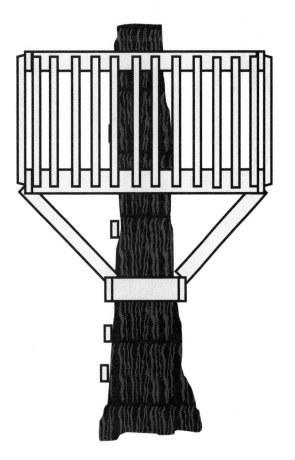

TOOLS REQUIRED

- Safety glasses
- 24" level
- Hand saw
- Gloves
- Hand drill
- $\frac{3}{16}$" drill bit
- Carpenter's hammer
- Screwdriver
- Paint brush
- 24" framing square
- Tape measure
- $\frac{9}{16}$" wrench

BILL OF MATERIAL

ITEM	QTY.	SIZE	DESCRIPTION	ITEM	QTY.	SIZE	DESCRIPTION
1	2	2 x 6 x 1'-6"	Lower support	11	10	$\frac{5}{4}$ x 6 x 20"	Deck
2	2	2 x 6 x 1'-9"	Lower support	12	2	2 x 4 x 5'-0"	Handrail
3	2	2 x 6 x 1'-6"	Upper inner support	13	2	2 x 4 x 5'-9"	Handrail
4	2	2 x 6 x 5'-9"	Upper inner support	14	46	2 x 2 x 34"	Guardrail posts
5	2	2 x 6 x 5'-0"	Perimeter frame	15	4	$\frac{5}{4}$ x 6 x 2'-6"	Bench
6	2	2 x 6 x 5'-9"	Perimeter frame	16	8	2 x 2 x 10"	Bench support
7	4	2 x 4 x 36"	Corner post	17	box	$\frac{3}{8}$" x 3 $\frac{1}{2}$" lag screws	Lag screws
8	8	2 x 4 x 36"	Gusset	18	box	1 $\frac{5}{8}$" deck screws	Deck screws
9	10	2 x 4 x 1'6"	Steps	19	box	3" deck screws	Deck screws
10	3	$\frac{5}{4}$ x 6 x 4'-11"	Deck	20	gal.	Optional	Paint

Simple Deck Tree House Plan

STEP 1 – INSTALL LOWER SUPPORT:

Locate a tree with a trunk diameter of approximately 18 inches minimum. The best trees for construction of a tree house are apple, ash, beech, cedar, Colorado blue spruce, cypress, fir, hemlock, hickory, sugar maple, oak, and weeping willow. Attach items 1 & 2, qty. two each to the tree trunk, using item 17 lag screws, qty. four per board. The elevation of the support should be 2 feet below the intended height of the tree house floor. As you install each board, install one screw and then check for level prior to installing the subsequent screws.

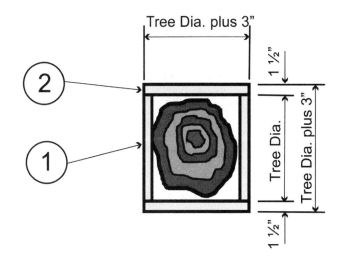

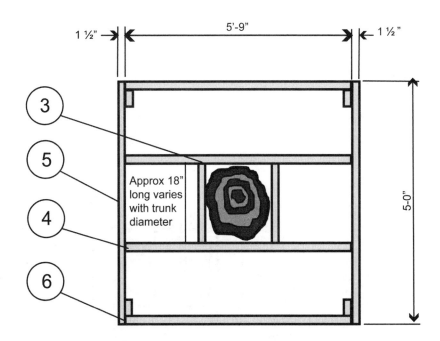

STEP 2 – INSTALL PERIMETER FRAME:

Attach inner supports items 3 & 4, qty. two each to the tree trunk, using item 17 lag screws, qty. four per board. The elevation of the support should be 2 feet above the lower support. As you install each board, install one screw and then check for level prior to installing the subsequent screws. Install the outer supports item 5 & 6, qty. two each, using item 19 lag screws, qty. two per connection. Install corner posts item 7, qty. four, flush with the bottom of the frame. Fasten securely using item 19 lag screws, qty. four per connection.

Simple Deck Tree House Plan

STEP 3 – INSTALL GUSSET SUPPORTS AND STEPS:

Cut item 8 gusset to approximate length of 36 inches. Notch one end and place notch on inside of upper perimeter frame. Align with outside edge of lower support and draw a line with a pencil to locate the lower notch. Attach item 8, qty. eight, to the upper and lower supports, using item 17 lag screws, qty. two at each connection. Check the upper platform with level prior to cutting the gussets. Fasten item 9 steps to tree using lag screws item 17, qty. four per step.

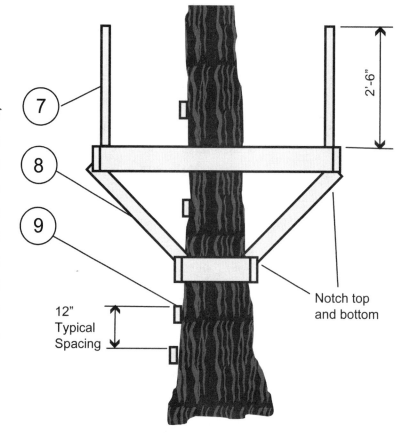

STEP 4 – INSTALL DECK:

Install deck item 10 & 11 using item 18 deck screws, 1 ⅝ inch long. Use 3 screws at each connection point. Space boards evenly leaving ¼-inch minimum gap to allow for expansion.

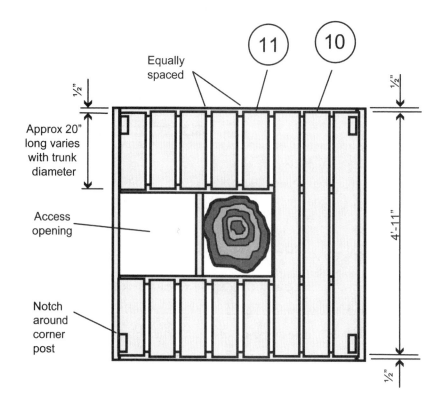

Simple Deck Tree House Plan

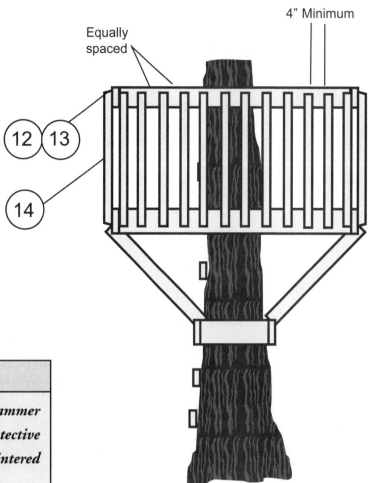

STEP 5 – INSTALL HANDRAIL:

Install top rail item 12 & 13 using item 19 deck screws, 3 inches long. Fasten item 14 side posts to top rail and perimeter frame using item 19 deck screws, 3 inches long. Space item 14 equally with a maximum gap of 4 inches.

CAUTION

Wear protective safety glasses when using hammer or operating power tools, and wear protective gloves when handling sharp objects and splintered wood.

STEP 6 – CORNER BENCH:

Attach bench support item 16, qty. eight to handrail posts using item 19 deck screws, 3 inches long. Install corner bench item 15, typical (4) places, using item 18 deck screws, 1 ⅝ inches long.

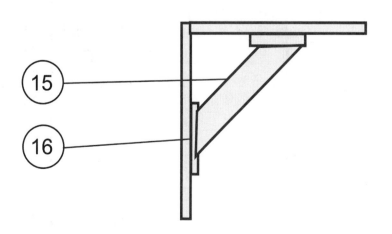

Western Fort Tree House Plan

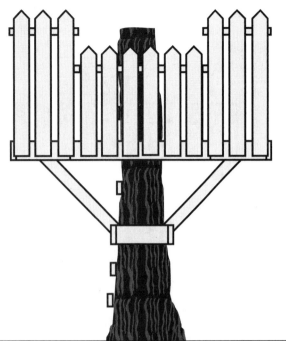

TOOLS REQUIRED

- Safety glasses
- Gloves
- Carpenter's hammer
- 24" framing square
- 24" level
- Hand drill

- Screwdriver
- Tape measure
- Hand saw
- $\frac{3}{16}$" drill bit
- Paint brush
- $\frac{9}{16}$" wrench

BILL OF MATERIAL

ITEM	QTY.	SIZE	DESCRIPTION	ITEM	QTY.	SIZE	DESCRIPTION
1	2	2 x 6 x 1'-6"	Lower support	16	16	2 x 4 x 1'9"	Tower floor support
2	2	2 x 6 x 1'-9"	Lower support	17	16	2 x 4 x 1'-0"	Tower floor support
3	2	2 x 6 x 1'-6"	Upper inner support	18	8	$\frac{5}{4}$ x 6 x 2'-10"	Tower floor
4	2	2 x 6 x 5'-9"	Upper inner support	19	8	$\frac{5}{4}$ x 6 x 1'-10"	Tower floor
5	2	2 x 6 x 7'-0"	Perimeter frame	20	2	2 x 2 x 2'-9"	Top rail
6	2	2 x 6 x 5'-9"	Perimeter frame	21	2	2 x 2 x 3'-9"	Top rail
7	8	2 x 6 x 12"	Corner tower frame	22	8	2 x 2 x 1'-3"	Top rail
8	4	2 x 6 x 21 ½"	Corner tower frame	23	4	2 x 2 x 2'-3"	Top rail
9	4	2 x 6 x 2'-3"	Corner tower frame	24	4	2 x 2 x 2'-1 ½"	Top rail
10	8	2 x 2 x 36"	Short corner post	25	28	$\frac{5}{4}$ x 6 x 3'-6"	Short wall
11	12	2 x 2 x 48"	Long corner post	26	48	$\frac{5}{4}$ x 6 x 4'-6"	Tall wall
12	10	2 x 4 x 1'-6"	Steps	27	box	$\frac{3}{8}$" x 3 ½" lag screws	Lag screws
13	8	2 x 4 x 3'-0"	Gusset	28	box	3" deck screws	Deck screws
14	3	$\frac{5}{4}$ x 6 x 4'-11"	Deck	29	box	1 $\frac{5}{8}$" deck screws	Deck screws
15	10	$\frac{5}{4}$ x 6 x 1'-8"	Deck	30	gal.	Optional	Paint

Western Fort Tree House Plan

STEP 1 – INSTALL LOWER SUPPORT:

Locate a tree with a trunk diameter of approximately 18 inches minimum. The best trees for construction of a tree house are apple, ash, beech, cedar, Colorado blue spruce, cypress, fir, hemlock, hickory, sugar maple, oak, and weeping willow. Attach items 1 & 2, qty. two each to the tree trunk, using item 27 lag screws, qty. four per board. The elevation of the support should be 2 feet below the intended height of the tree house floor. As you install each board, install one screw and check for level prior to installing the subsequent screws.

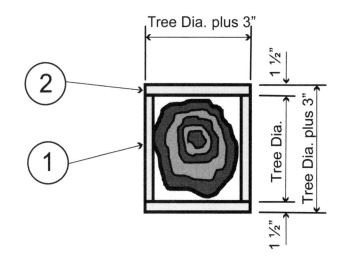

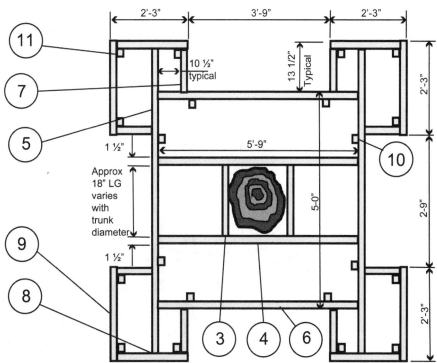

STEP 2 – INSTALL PERIMETER FRAME:

Attach inner supports items 3 & 4, qty. two each to the tree trunk, using item 27 lag screws, qty. four per board. The elevation of the support should be 2 feet above the lower support. As you install each board, install one screw and check for level prior to installing the subsequent screws. Install the outer supports items 5 & 6, qty. two each, using item 27 lag screws, qty. two per connection. Assemble the corner tower supports items 7, 8, & 9 using lag screws item 27, qty. four per connection. Install corner posts items 10 & 11, flush with the bottom of the frame. **NOTE**: Long posts are in the corner tower. Fasten securely using item 27 lag screws, qty. two per connection.

Western Fort Tree House Plan

STEP 3 – INSTALL GUSSET SUPPORTS AND STEPS:

Cut item 13 gusset to approximate length of 36 inches. Notch one end; place notch on inside of upper perimeter frame. Align with outside edge of lower support and draw a line with a pencil to locate the lower notch. Attach item 13, qty. eight to the upper and lower supports, using item 27 lag screws, qty. two at each connection. Check the upper platform with level prior to cutting the gussets. Fasten item 12 steps to tree using lag screws item 27, qty. four per step.

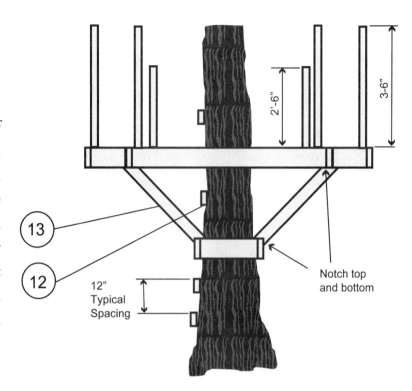

STEP 4 – INSTALL MAIN DECK:

Install deck items 14 & 15 using item 29 deck screws, 1 ⅝ inches long. Use three screws at each connection point. Space boards evenly, leaving ¼ inches minimum gap to allow for expansion.

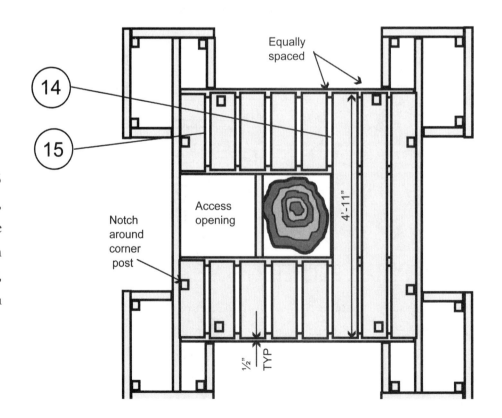

Western Fort Tree House Plan

STEP 5 – INSTALL FORT WALLS:

Install top rail items 20, 21, 22, 23, & 24 using item 28 deck screws, 3 inches long. Fasten items 25 & 26 side walls to top rail and perimeter frame using item 29 deck screws, 1 ⅝ inches long. Space equally.

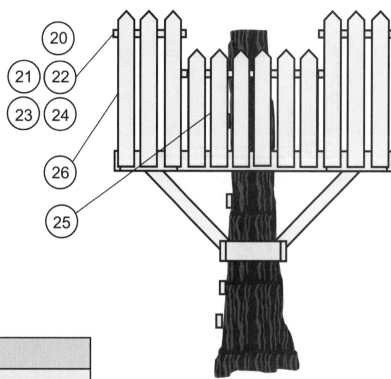

<table>
<tr><td colspan="2" align="center">CAUTION</td></tr>
<tr><td colspan="2">Wear protective safety glasses when using hammer or operating power tools, and wear protective gloves when handling sharp objects and splintered wood.</td></tr>
</table>

STEP 6 – INSTALL CORNER TOWER PLATFORM:

Attach corner tower deck supports item 16 & 17, qty. eight, to corner posts using item 28 deck screws, 3 inches long. Corner tower floor will be located 12 inches above main deck. Install corner tower floor item 18 & 19, using item 29 deck screws, 1 ⅝ inches long.

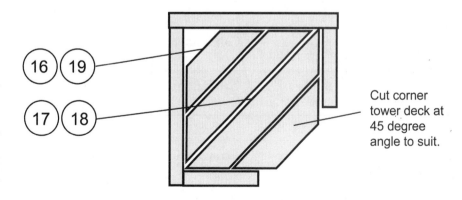

Cut corner tower deck at 45 degree angle to suit.

Kiddie Cabin Tree House Plan

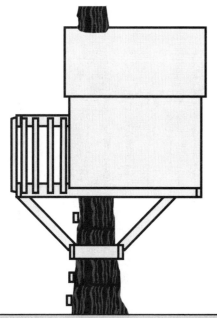

TOOLS REQUIRED

• Safety glasses	• Screwdriver
• Gloves	• Tape measure
• Carpenter's hammer	• Hand saw
• 24" framing Square	• $\frac{3}{16}$" drill bit
• 24" level	• Paint brush
• Hand drill	• $\frac{9}{16}$" wrench

BILL OF MATERIAL

ITEM	QTY.	SIZE	DESCRIPTION	ITEM	QTY.	SIZE	DESCRIPTION
1	2	2 x 6 x 1'-6"	Lower support	17	6	2 x 2 x 3'-8"	Wall support
2	2	2 x 6 x 1'-9"	Lower support	18	4	2 x 2 x 3'-4"	Roof support
3	3	2 x 8 x 1'-6"	Upper inner support	19	2	2 x 2 x 6'-2"	Wall support
4	2	2 x 8 x 6'-9"	Upper inner support	20	1	2 x 2 x 4'-10 ½"	Roof peak support
5	2	2 x 8 x 5'-0"	Perimeter frame	21	2	½" ply x 3'-6" x 5'-2 ½"	Roof
6	2	2 x 8 x 6'-9"	Perimeter frame	22	2	2 x 4 x 1'-11 ½"	Handrail
7	4	2 x 4 x 36"	Corner post	23	1	2 x 4 x 4'-9"	Handrail
8	8	2 x 4 x 36"	Gusset	24	18	2 x 2 x 3'-0"	Guardrail posts
9	10	2 x 4 x 1'6"	Steps	25	4	½" ply x 8" x 12"	Shutters
10	1	¾ ply x 4'-0" x 4'-11"	Cabin floor	26	1	½" ply x 1'-6" x 3'-0"	Door
11	1	¾ ply x 9 ½" x 4'-11"	Cabin floor	27	10	2" hinge	Hinge
12	6	$\frac{5}{4}$ x 6" x 20"	Deck	28	box	$\frac{3}{8}$" x 3 ½" lag screws	Lag screws
13	1	½" ply x 5'-0" x 6'-6"	Front wall	29	box	1 $\frac{5}{8}$" deck screws	Deck screws
14	1	½" ply x 5'-0" x 6'-6"	Back wall	30	box	3" deck screws	Deck screws
15	2	½" ply x 4'-0" x 4'-10 ½"	Side wall	31	gal.	Optional	Paint
16	4	2 x 2 x 4'-10 ½"	Wall support	32	bdl.	Standard size	Shingles

Kiddie Cabin Tree House Plan

STEP 1 – INSTALL LOWER SUPPORT:

Locate a tree with a trunk diameter of approximately 18 inches minimum. The best trees for construction of a tree house are apple, ash, beech, cedar, Colorado blue spruce, cypress, fir, hemlock, hickory, sugar maple, oak, and weeping willow. Attach items 1 & 2, qty. two each, to the tree trunk, using item 28 lag screws, qty. four per board. The elevation of the support should be 2 feet below the intended height of the tree house floor. As you install each board, install one screw and then check for level prior to installing the subsequent screws.

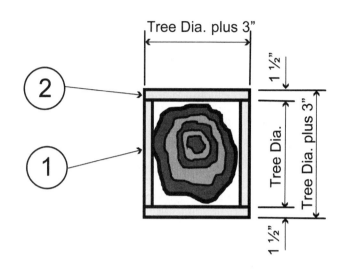

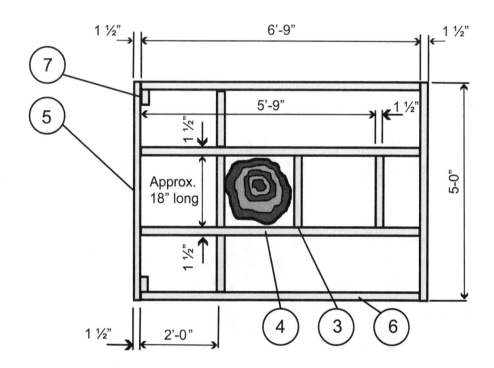

STEP 2 – INSTALL PERIMETER FRAME:

Attach inner supports items 3 & 4 to the tree trunk, using item 28 lag screws, qty. four per board. The elevation of the support should be 2 feet above the lower support. As you install each board, install one screw and then check for level prior to installing the subsequent screws. Install the outer supports items 5 & 6, qty. two each, using item 28 lag screws, qty. two per connection. Install corner posts item 7, flush with the bottom of the frame. Fasten securely using item 28 lag screws, qty. four per connection.

Kiddie Cabin Tree House Plan

STEP 3 – INSTALL GUSSET SUPPORTS AND STEPS:

Cut item 8 gusset to approximate length of 36 inches. Notch one end and place notch on inside of upper perimeter frame. Align with outside edge of lower support and draw a line with a pencil to locate the lower notch. Attach item 8, qty. eight, to the upper and lower supports, using item 28 lag screws, qty. two at each connection. Check the upper platform with level prior to cutting the gussets. Fasten item 9 steps to tree using lag screws item 28, qty. four per step.

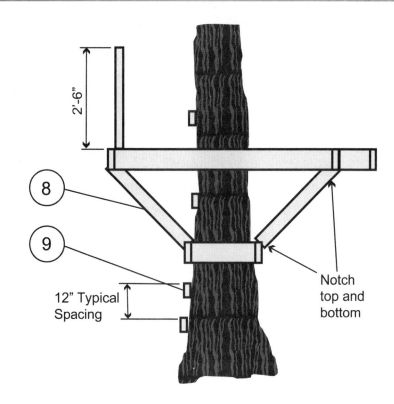

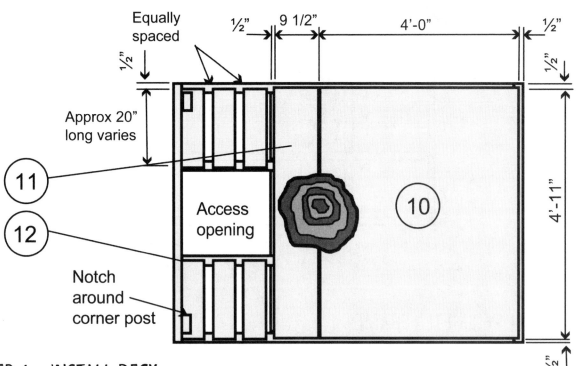

STEP 4 – INSTALL DECK:

Install deck items 10, 11, & 12 using item 29 deck screws, 1 ⅝ inches long. Use three screws at each connection point. Space boards evenly leaving ¼-inch minimum gap to allow for expansion. Notch plywood deck to match tree profile, leaving a 1-inch gap for clearance.

Kiddie Cabin Tree House Plan

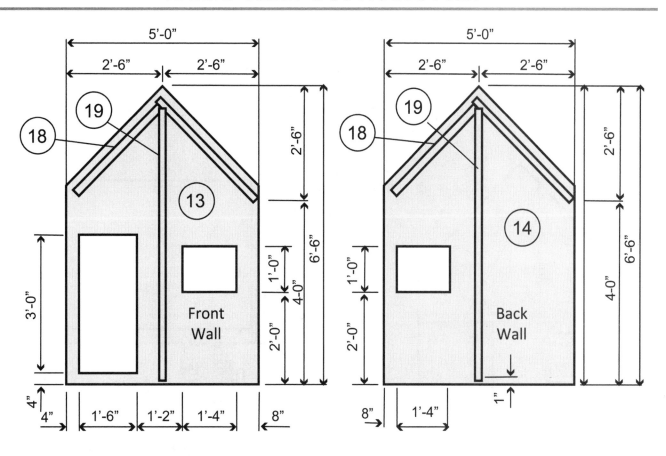

STEP 5 – ASSEMBLE CABIN WALLS:

Cut plywood as shown for front wall item 13 and back wall item 14. Save cutouts to use as shutters item 25 and door item 26. Attach supports items 18 & 19 to wall as shown using 1 ⅝-inch deck screws item 29.

Cut item 15 plywood as shown. Attach items 16 & 17 to side wall using 1 ⅝-inch deck screws item 29.

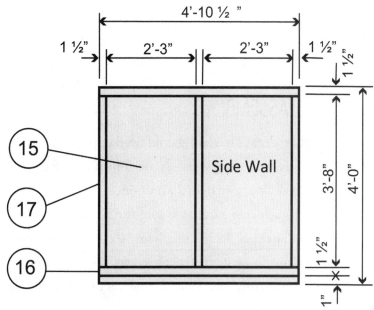

Kiddie Cabin Tree House Plan

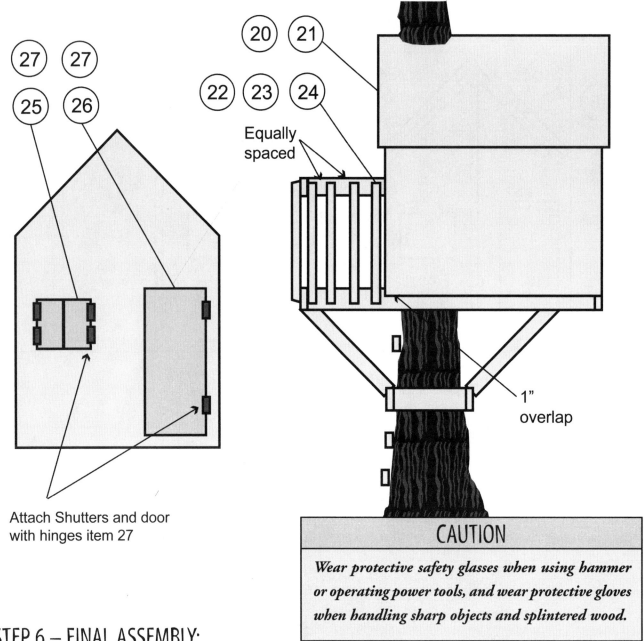

Attach Shutters and door with hinges item 27

Equally spaced

1" overlap

CAUTION
Wear protective safety glasses when using hammer or operating power tools, and wear protective gloves when handling sharp objects and splintered wood.

STEP 6 – FINAL ASSEMBLY:

Assemble back wall to deck. Plywood should overlap perimeter frame by 1 inch. Screw plywood to perimeter frame using 1 5/8-inch deck screws item 29 on 6-inch spacing. Attach side walls to perimeter frame and end wall using 1 5/8-inch deck screws item 29 on 6-inch spacing. Install front wall assembly using 1 5/8-inch deck screws item 29 on 6-inch centers. Attach roof sheet item 21 to end walls and side walls with 1 5/8-inch deck screws. Install top rail item 22 & 23 using item 30 deck screws, 3 inches long. Fasten item 24 side posts to top rail and perimeter frame using item 30 deck screws, 3 inches long. Space item 24 equally with a maximum gap of 4 inches. Use cutouts from end walls to construct hinged door item 26 and hinged shutters item 25, and attach to end wall with hinges item 27.

Pirate Ship Tree House Plan

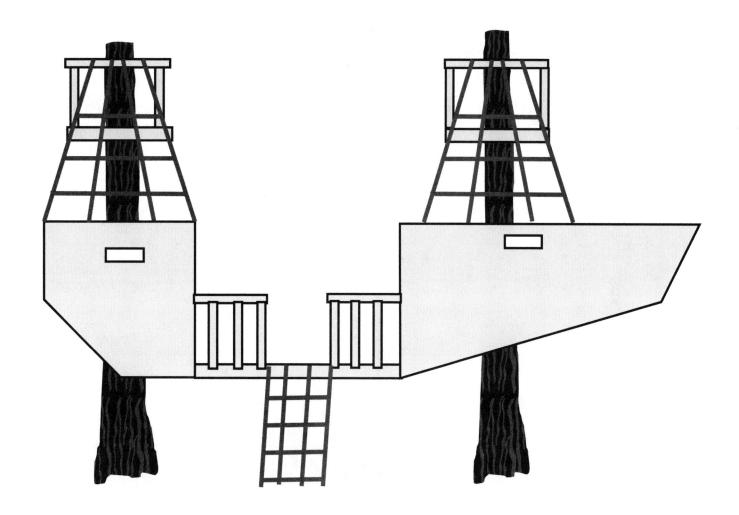

TOOLS REQUIRED	
• Safety glasses	• Screwdriver
• Gloves	• Tape measure
• Carpenter's hammer	• Hand saw
• 24" framing square	• $\frac{3}{16}$" drill bit
• 24" level	• Paint brush
• Hand drill	• $\frac{9}{16}$" wrench

Pirate Ship Tree House Plan

ITEM	QTY.	SIZE	DESCRIPTION	ITEM	QTY.	SIZE	DESCRIPTION
1	7	2 x 6 x 1'-6"	Support	17	2	2 x 2 x 2'-6"	Rail support
2	4	2 x 6 x 1'-9"	Support	18	2	2 x 2 x 8'-0"	Rail support
3	4	2 x 6 x 3'-9"	Deck support	19	2	½" ply x 4'-0" x 5'-0"	Stern side wall
4	4	2 x 6 x 4'-0"	Deck support	20	2	½" ply x 5'-0" x 8'-0"(splice)	Bow side wall
5	60	2 x 2 x 3'-6"	Rail supports	21	1	½" ply x 3'-6" x 4'-0"	Stern rear wall
6	4	4 x 4 x 5'-6"	Mid-deck hanger	22	4	2 x 4 x 1'-6"	Crow support
7	14	2 x 4 x 3'-0"	Gusset	23	8	2 x 4 x 2'-6"	Crow support
8	4	2 x 10 x 8'-0"	Support	24	4	2 x 4 x 2'-9"	Crow support
9	3	2 x 10 x 1'-2"	Support	25	4	2 x 2 x 2'-6"	Crow handrail
10	18	2 x 10 x 4'-0"	Deck	26	4	2 x 2 x 2'-9"	Crow handrail
11	6	⁵⁄₄ x 6 x 14"	Deck	27	8	2 x 4 x 1'-6"	Steps
12	14	⁵⁄₄ x 6 x 3'-11"	Deck	28	box	½" x 4" lag screws	Lag screws
13	10	⁵⁄₄ x 6 cut to suit	Deck	29	box	1" deck screws	Deck screws
14	2	2 x 2 x 3'-9"	Rail supports	30	box	1 ⁵⁄₈" deck screws	Deck screws
15	1	2 x 2 x 4'-0"	Rail support	31	box	3" deck screws	Deck screws
16	1	2 x 2 x 7'-6"	Rail support	32	100'	½" rope cut to suit	Rope

Pirate Ship Tree House Plan

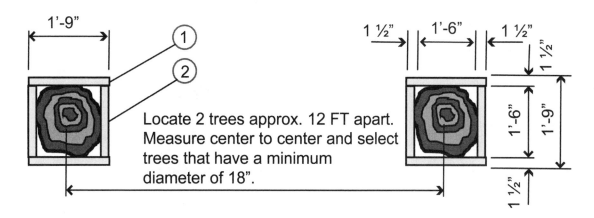

Locate 2 trees approx. 12 FT apart. Measure center to center and select trees that have a minimum diameter of 18".

STEP 1 – INSTALL LOWER SUPPORTS:

The best trees for the construction of a tree house are apple, ash, beech, cedar, Colorado blue spruce, cypress, fir, hemlock, hickory, sugar maple, oak, and weeping willow. Attach items 1 & 2, qty. four each to the tree trunk using item 28 lag screws, qty. four per board. Use a ¼-inch drill bit and drill pilot holes to prevent wood from splitting. The elevation of the support should be 2 feet below the intended height of the tree house's upper floor. As you install each board, install one screw and then check for level prior to installing the subsequent screws.

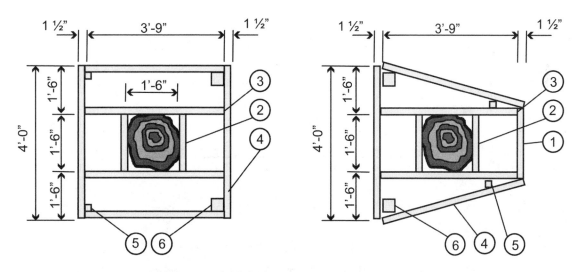

STEP 2 – CONSTRUCT STERN & BOW UPPER DECKS:

Attach supports items 2 & 3 to tree using item 28 lag screws, qty. four per connection. Assemble perimeter supports, using two lag screws per connection. Install corner supports item 5 flush with the bottom of the frame and attach using deck screws item 28, qty. four screws per connection. Install mid-deck hanger item 6 protruding 2 feet - 6 inches below bottom of frame. Attach to outer frame using qty. eight lag screws item 28 per connection. Notch gusset item 7 as shown below and install between upper deck and lower supports typical (14) places using lag screws.

Pirate Ship Tree House Plan

STEP 3 – INSTALL GUSSETS:

Notch gussets item 7 to mate with upper perimeter frame and lower support. Attach item 11 using lag screws item 28, qty. two per connection.

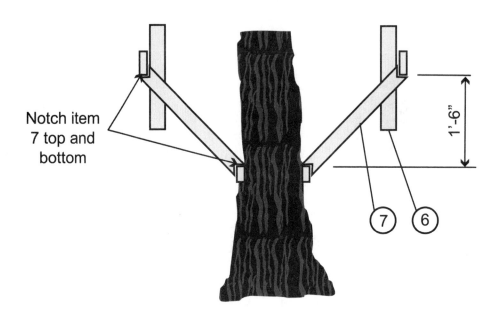

Notch item 7 top and bottom

STEP 4 –ASSEMBLE MAIN DECK:

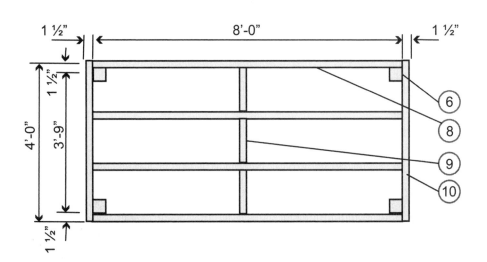

Attach items 4 & 8 to item 6 hanger using eight lag screws item 28 per connection. Bottom of main deck will be flush with the bottom of item 6. Assemble remaining components item 8 & 9 using two lag screws item 28 per connection.

Pirate Ship Tree House Plan

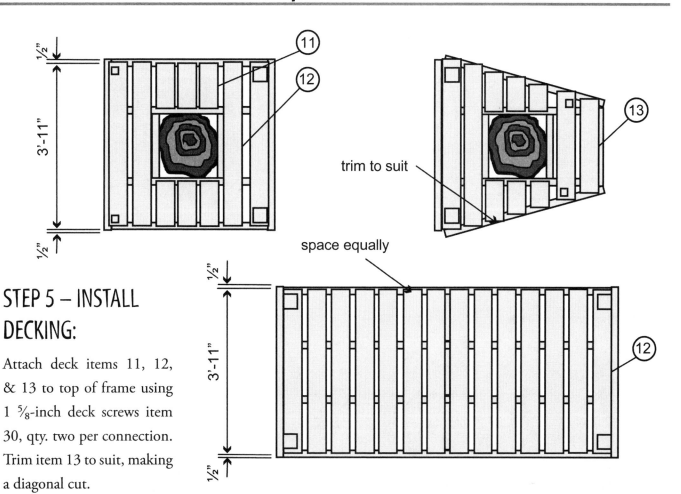

STEP 5 – INSTALL DECKING:

Attach deck items 11, 12, & 13 to top of frame using 1 ⅝-inch deck screws item 30, qty. two per connection. Trim item 13 to suit, making a diagonal cut.

trim to suit

space equally

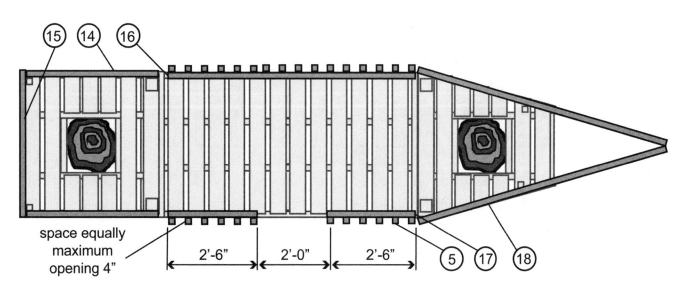

space equally maximum opening 4"

2'-6"   2'-0"   2'-6"

STEP 6 – INSTALL HANDRAIL AND POSTS:

Attach vertical posts item 5 to outside of frame using 3-inch deck screws item 31, qty. three screws per connection. Install top handrail items 14, 15, 16, 17, & 18 using 3-inch deck screws item 31, qty. two per connection.

Pirate Ship Tree House Plan

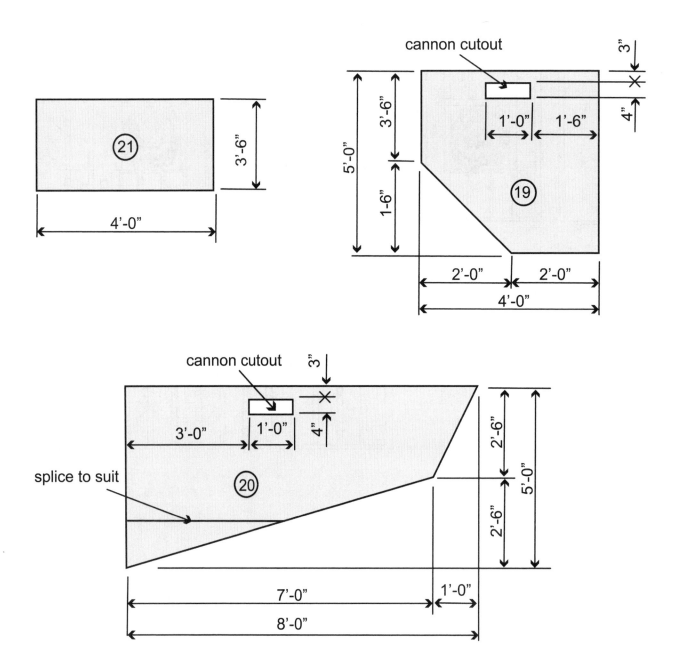

STEP 7- PREPARE PLYWOOD WALLS:

Cut out plywood walls items 19, 20, & 21 as shown. Attach plywood to perimeter frame of upper decks using 1 ⅝-inch deck screws item 30 on 6-inch spacing. Plywood will protrude 3 feet - 0 inches above top of deck. Pull front of bow together at the bottom to form a slight curve and fasten with 1-inch deck screws item 29.

Pirate Ship Tree House Plan

STEP 8 – CONSTRUCT CROW'S NEST:

Assemble crow's nest 4 feet above upper decks. Attach items 22 & 23 to tree using lag screws item 28, qty. four per connection. Assemble outer perimeter items 23 & 24 using lag screws. Install handrail items 25 & 26 and supports item 5 using 3-inch deck screws item 31, qty. three screws per connection.

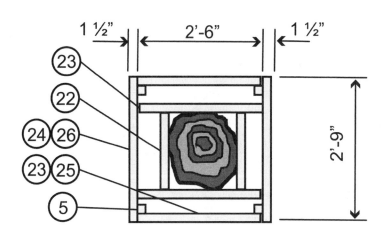

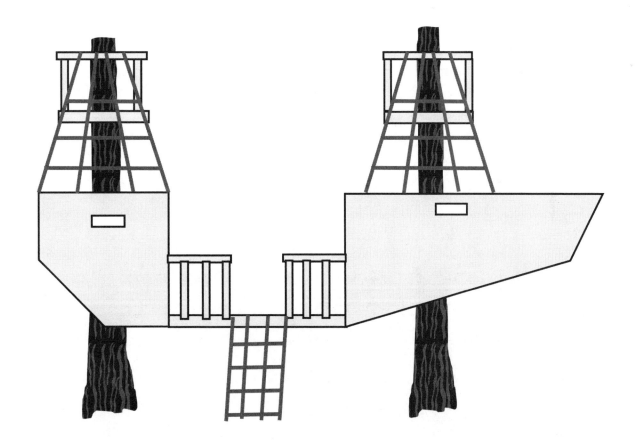

STEP 9 – ASSEMBLE ROPE MESH:

Assemble rope mesh for access to main deck and crow's nest.

CAUTION

Wear protective safety glasses when using hammer or operating power tools, and wear protective gloves when handling sharp objects and splintered wood.

Bird Watcher's Paradise Tree House Plan

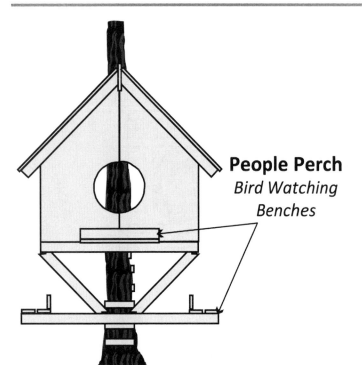

People Perch
Bird Watching Benches

TOOLS REQUIRED

• Safety glasses	• Screwdriver
• Gloves	• Tape measure
• Carpenter's hammer	• Hand saw
• 24" framing square	• ¼" drill bit
• 24" level	• Paint brush
• Hand drill	• ¾" Wrench

BILL OF MATERIAL

ITEM	QTY.	SIZE	DESCRIPTION	ITEM	QTY.	SIZE	DESCRIPTION
1	2	2 x 8 x 10'-0"	Lower perch	16	2	½" ply x 4'-0" x 8'-0"	Side wall
2	10	2 x 8 x 2'-0"	Perch support	17	4	½" ply x 4'-0" x 8'-0"	End wall
3	12	2 x 8 x 4'-0"	Perch bench	18	1	2 x 10 x 10'-0"	Roof truss top
4	6	⁵⁄₄ x 6 x 2'-2"	Perch deck	19	10	2 x 6 x 7'-0"	Roof truss
5	2	2 x 10 x 12'-0"	Upper perch	20	4	2 x 4 x 6'-0 ½"	Wall support
6	6	2 x 10 x 2'-9"	Deck support	21	4	2 x 4 x 2'-6"	Wall support
7	1	2 x 10 x 7'-9"	Deck support	22	4	2 x 4 x 4'-0"	Wall support
8	4	2 x 10 x 1'-3 ¾"	Deck support	23	2	2 x 4 x 8'-0"	Wall bottom board
9	2	2 x 10 x 8'-0"	Deck perimeter	24	4	½" ply x 4'-0" x 8'-0"	Roof deck
10	10	2 x 4 x 1'-6"	Steps	25	1 bdl.	Standard roof size	Shingles
11	8	2 x 6 x 5'-0"	Gusset	26	box	½" x 4" lag screws	Lag screws
12	10	⁵⁄₄ x 6 x 8'-0"	Deck	27	box	1" roofing nails	Nails
13	4	⁵⁄₄ x 6 x 3'-0"	Deck	28	box	1 ⁵⁄₈" deck screws	Deck screws
14	4	2 x 4 x 7'-5"	Side wall frame	29	box	3" deck screws	Deck screws
15	10	2 x 4 x 3'-9"	Side wall frame				

Bird Watcher's Paradise Tree House Plan

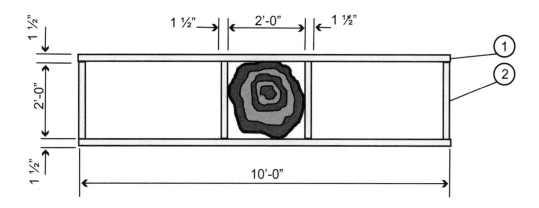

STEP 1 – INSTALL LOWER "PEOPLE PERCH" SUPPORTS:

The best trees for the construction of a tree house are apple, ash, beech, cedar, Colorado blue spruce, cypress, fir, hemlock, hickory, sugar maple, oak, and weeping willow. Attach items 1 & 2, qty. two each to the tree trunk, using item 26 lag screws, qty. four per connection. Use a ¼-inch drill bit and drill pilot holes to prevent wood from splitting. The elevation of the support should be 4 feet below the intended height of the tree house's upper floor. As you install each board, install one screw and then check for level prior to installing the subsequent screws. Assemble additional perimeter supports item 2 using lag screws item 26.

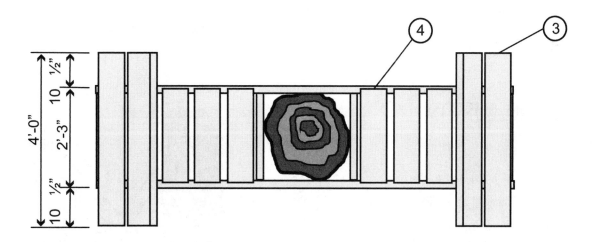

STEP 2 – CONSTRUCT LOWER "PEOPLE PERCH" BENCHES:

Attach perch benches item 3 to top of perch supports using 3-inch deck screws item 29 using 3 screws per connection. Fasten deck boards item 4 to deck using 1 ⅝-inch deck screws item 28, qty. three per connection.

Bird Watcher's Paradise Tree House Plan

STEP 3 – ASSEMBLE UPPER "PEOPLE PERCH":

Attach upper perch items 2 & 5 to tree using lag screws item 26, qty. six per connection. Assemble perimeter supports items 2, 6, 7, & 9 using lag screws item 26, qty. three per connection.

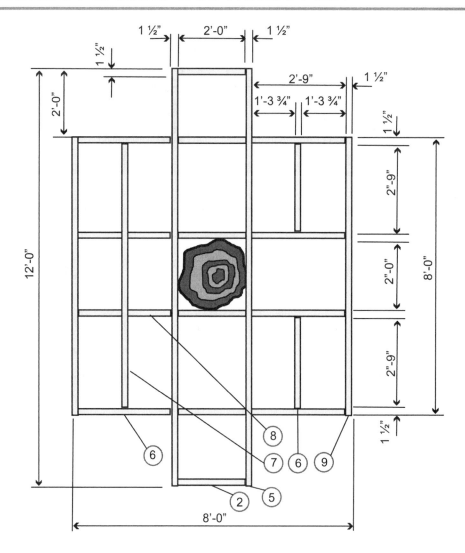

STEP 4 – INSTALL GUSSETS & STEPS:

Notch gussets item 11 to mate with upper perimeter frame and lower support. Attach item 11 using lag screws item 26, qty. two per connection. Install steps item 12 using lag screws item 26, qty. four per connection. Lower steps are located on the front of the tree. Upper steps are located on the right side of the tree.

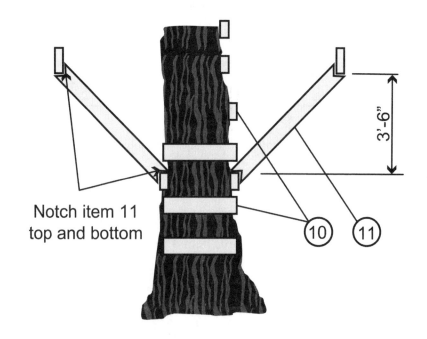

Notch item 11 top and bottom

Bird Watcher's Paradise Tree House Plan

STEP 5 – INSTALL DECK AND UPPER "PEOPLE PERCH":

Attach perch item 3 to top of perch support using 3-inch deck screws item 29, qty. three per connection. Fasten deck items 12 & 13 to frame using 1 ⅝-inch deck screws item 28, qty. three per connection.

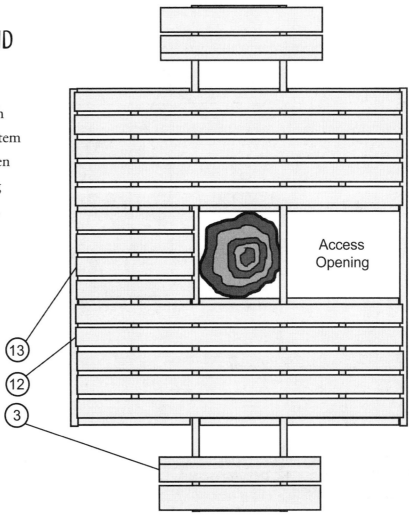

Access Opening

(13)

(12)

(3)

STEP 6 – ASSEMBLE SIDE WALLS, QTY. TWO:

Assemble items 14 & 15 frame using 3-inch deck screws item 29, qty. two per connection. Attach wall item 16 to frame assembly using 1 ⅝-inch deck screws item 28 on 6-inch spacing.

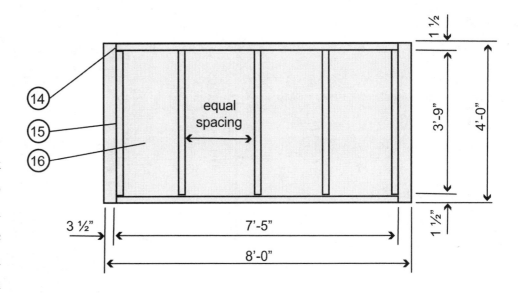

(14)

(15)

(16)

equal spacing

1 ½

3'-9"

4'-0"

1 ½"

3 ½"

7'-5"

8'-0"

Bird Watcher's Paradise Tree House Plan

STEP 7 – CUT OUT PLYWOOD END WALLS:

Cut out item 17 from ½-inch exterior-use treated plywood. A total qty. of four required.

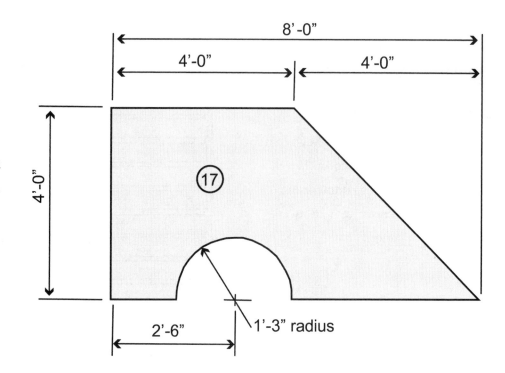

STEP 8 – ASSEMBLE END WALLS:

Assemble end walls qty. two, including items 17, 18, 19, 20 , 21, 22, & 23. Fasten frame together using 3-inch deck screws item 29. Attach plywood to frame using 1 ⅝-inch deck screws item 28. Attach assembly to top surface of deck using lag bolts item 26 on 6-inch spacing.

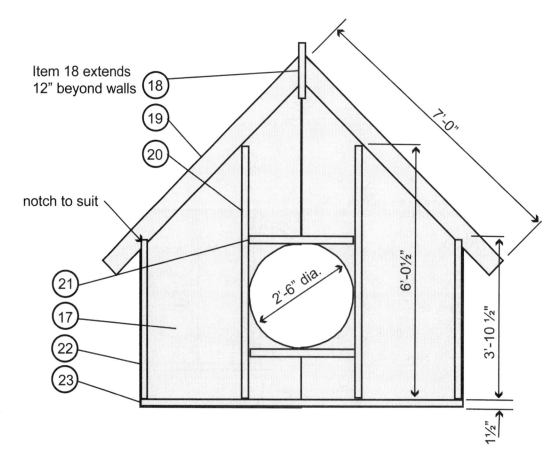

Bird Watcher's Paradise Tree House Plan

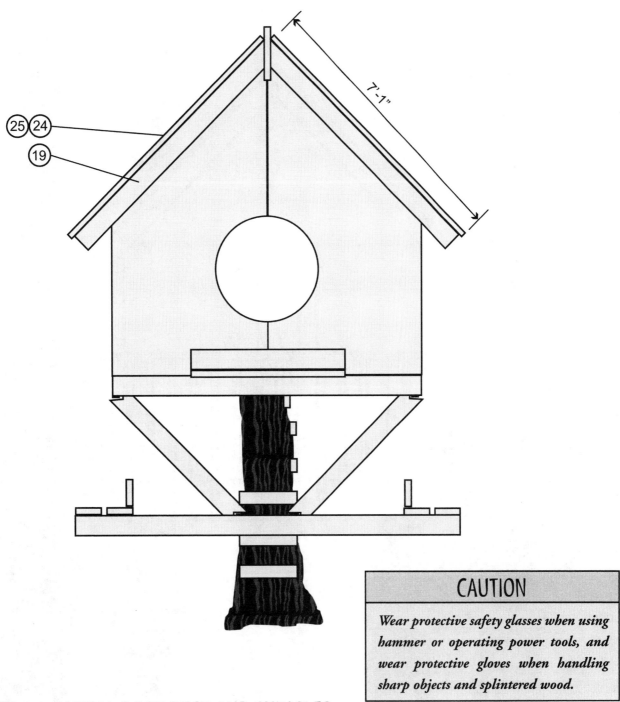

CAUTION

Wear protective safety glasses when using hammer or operating power tools, and wear protective gloves when handling sharp objects and splintered wood.

STEP 9 – INSTALL ROOF DECK AND SHINGLES:

Install remaining roof trusses item 19 on 24-inch centers using 3-inch deck screws item 29, qty. two per connection. Attach roof deck item 24 to side walls and end walls using 3-inch long deck screws item 29, qty. three per connection. Install shingles item 25 per manufacturer's recommended nailing pattern.

Cabin in the Sky With Wrap-Around Deck House Plan

Cabin in the Sky With Wrap-Around Deck House Plan

TOOLS REQUIRED

- Safety glasses
- Gloves
- Carpenter's hammer
- 24" framing square
- 24" level
- Hand drill
- Screwdriver
- Tape measure
- Hand saw
- $\frac{3}{16}$" drill bit
- Paint brush
- $\frac{3}{4}$" wrench

BILL OF MATERIAL

ITEM	QTY.	SIZE	DESCRIPTION	ITEM	QTY.	SIZE	DESCRIPTION
1	8	2 x 6 x 2'-0"	Lower support	22	2	2 x 4 x 9'-5"	Wall base
2	8	2 x 6 x 2'-3"	Lower support	23	1	2 x 12 x 12'-0"	Roof truss
3	7	2 x 12 x 14'-0"	Deck frame	24	16	2 x 8 x 7'-0"	Roof truss
4	9	2 x 12 x 1'-10 ½"	Deck frame	25	2	2 x 4 x 2'-0"	Wall frame
5	13	2 x 12 x 2'-0"	Deck frame	26	4	2 x 4 x 8'-4 ½"	Wall frame
6	9	2 x 12 x 3'-0"	Deck frame	27	4	2 x 4 x 7'-1 ½"	Wall frame
7	1	2 x 12 x 3'-6"	Deck frame	28	4	2 x 4 x 5'-10 ½"	Wall frame
8	7	2 x 12 x 7'-6"	Deck frame	29	4	2 x 4 x 3'-0"	Wall base
9	6	2 x 12 x 1'-2 ¾"	Deck frame	30	4	$\frac{5}{4}$ x 6 x 1'-11 ½"	Door frame
10	10	2 x 4 x 1'-6"	Steps	31	2	½" ply x 1'-11 ½" x 5'-11 ½"	Door
11	20	2 x 6 x 3'-6"	Gusset	32	2	$\frac{5}{4}$ x 6 x 5'-0 ½"	Door frame
12	5	$\frac{5}{4}$ x 6 x 9'-10 ½"	Deck	33	1	$\frac{5}{4}$ x 6 x 5'-2"	Door frame
13	3	$\frac{5}{4}$ x 6 x 14'-2"	Deck	34	2	Standard door latch	Latch
14	2	½" ply x 4'-0" x 8'-0"	Deck	35	4	2" hinges	Hinge
15	1	½" ply x 2'-0" x 8'-0"	Deck	36	box	½ x 4" lag screws	Screw
16	5	$\frac{5}{4}$ x 6 x 7'-10 ½"	Deck	37	box	1" deck screws	Screw
17	3	$\frac{5}{4}$ x 6 x 9'-8"	Deck	38	box	1 $\frac{5}{8}$" deck screws	Screw
18	130	2 x 2 x 4'-0"	Post	39	box	3" deck screws	Screw
19	8	2 x 4 x 14'-0"	Handrail	40	box	1" roofing nails	Nails
20	16	½" exterior ply, cut to suit	Walls, roof	41	2bd	Standard roof size	Shingles
21	16	2 x 4 x 5'-9"	Wall frame	42	gal.	Optional	Paint

Cabin in the Sky With Wrap-Around Deck House Plan

STEP 1 – ASSEMBLE LOWER-SUPPORT FRAMES:

The best trees for the construction of a tree house are apple, ash, beech, cedar, Colorado blue spruce, cypress, fir, hemlock, hickory, sugar maple, oak, and weeping willow. Attach items 1 & 2, qty. eight each to the tree trunk using item 36 lag screws qty. four per board. Use a ¼-inch drill bit and drill pilot holes to prevent wood from splitting. The elevation of the support should be 2 feet below the intended height of the tree house's main upper deck. As you install each board, install one screw and then check for level prior to installing the subsequent screws.

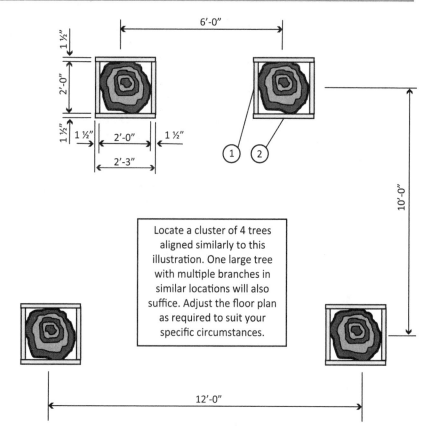

Locate a cluster of 4 trees aligned similarly to this illustration. One large tree with multiple branches in similar locations will also suffice. Adjust the floor plan as required to suit your specific circumstances.

STEP 2 – ASSEMBLE MAIN SUPPORT FRAME:

Assemble the main frame by first attaching item 3 to the tree trunks at the desired elevation. Fasten securely using item 36 lag screws, qty. eight per connection. Drill a pilot hole using a ¼-inch drill bit to prevent the wood from splitting. As you begin, confirm the first beam item 3 is level. All subsequent beams must be installed at the same elevation. Continue to assemble the remaining items 3, 4, 5, 6, 7, 8, & 9 using lag screws item 36, qty. four per connection.

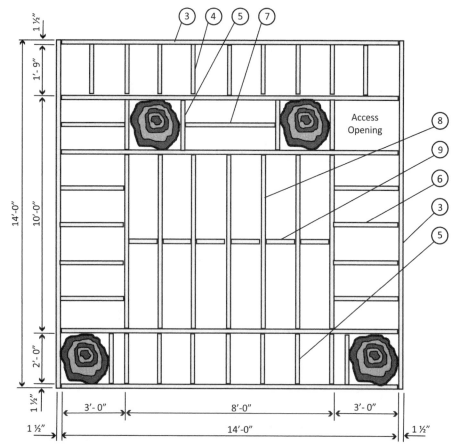

Cabin in the Sky With Wrap-Around Deck House Plan

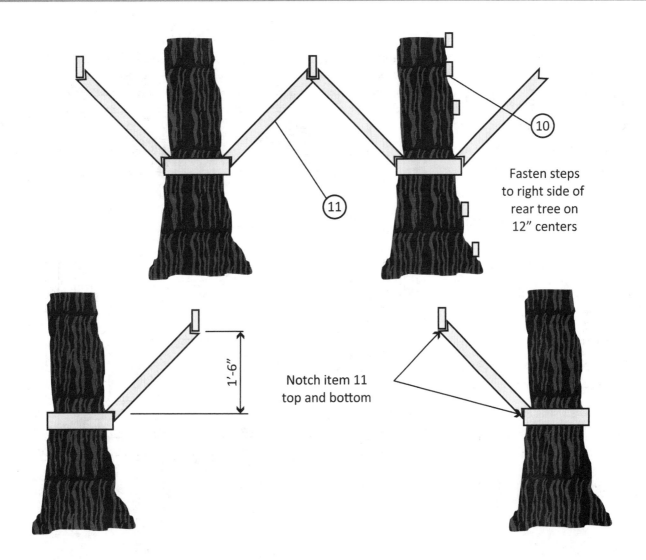

Fasten steps
to right side of
rear tree on
12" centers

1'-6"

Notch item 11
top and bottom

STEP 3 – INSTALL GUSSET SUPPORTS FOR MAIN FRAME:

Notch item 11 gusset at top and bottom to match lower support and upper main deck.
Fasten to frame using item 36 lag bolts qty. four per connection. Fasten steps item 10 to
right side of right rear tree using item 36 lag bolts, qty. four per connection.

Cabin in the Sky With Wrap-Around Deck House Plan

STEP 4 – INSTALL MAIN DECK:

Assemble the main deck. Attach plywood deck to frame using 1 ⅝-inch deck screws item 38 on 6-inch spacing. Attach perimeter deck to frame using 1 ⅝-inch deck screws item 38, qty. three per connection. Install railing supports item 18 around perimeter of deck using 3-inch deck screws item 39, qty. four per connection. Maximum spacing of item 18 is 4 inches. Fasten railing item 19 to top and inside of posts using 3-inch deck screws item 39, qty. two per connection.

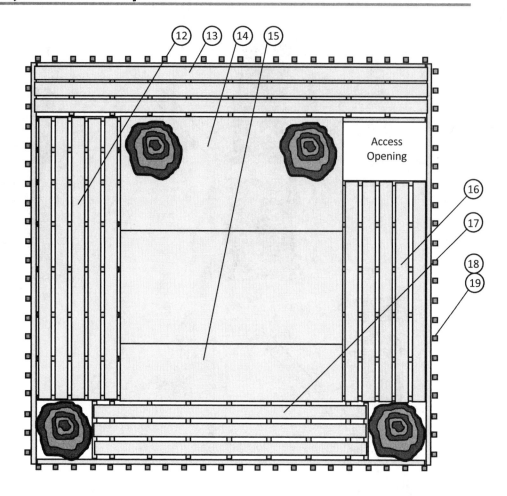

STEP 5 – ASSEMBLE SIDE WALLS:

Splice plywood item 20 to match overall dimensions as shown. Attach plywood to frame items 21 & 22 using 1 ⅝-inch deck screws on 6-inch spacing.

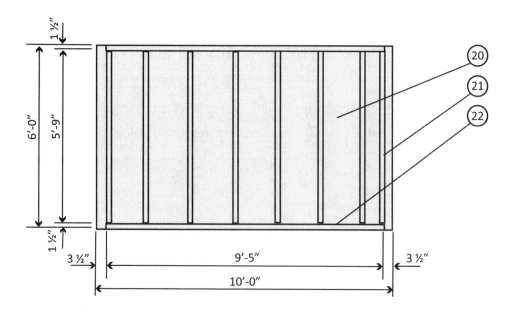

Cabin in the Sky With Wrap-Around Deck House Plan

STEP 6 – CUT OUT END-WALL DETAILS:

Cut out plywood end walls item 20 to match overall dimensions as shown.

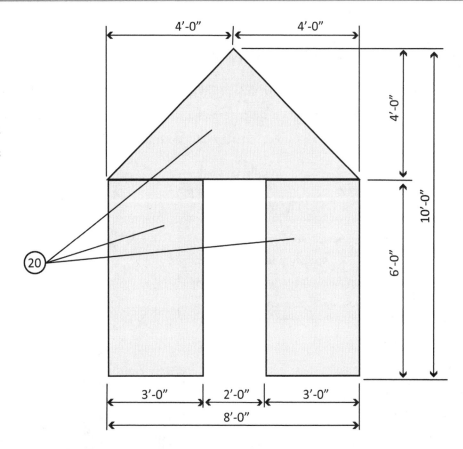

STEP 7 – ASSEMBLE END-WALL FRAME:

Assemble plywood end wall and end wall frame items 24, 25, 26, 27, 28, & 29 using 1 ⅝-inch deck screws item 38 on 6-inch spacing.

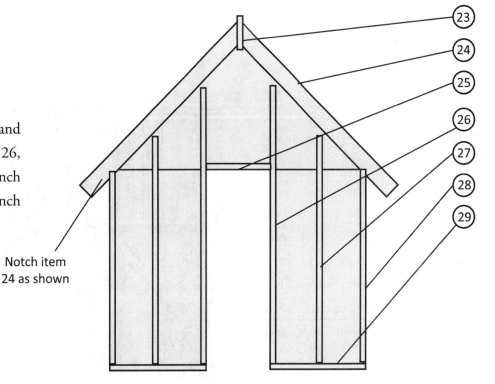

Notch item 24 as shown

Cabin in the Sky With Wrap-Around Deck House Plan

STEP 8 – ASSEMBLE HINGED DOORS:

Attach door frame items 30, 32, & 33 to plywood door item 31 using 1-inch deck screws item 37 on 6-inch centers. Install hinges item 35 and latch item 34.

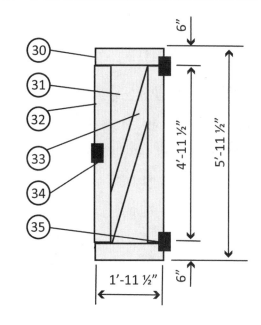

> ### CAUTION
>
> *Wear protective safety glasses when using hammer or operating power tools, and wear protective gloves when handling sharp objects and splintered wood.*

STEP 9 – ATTACH WALLS TO MAIN DECK AND INSTALL ROOF DECK AND SHINGLES:

Attach walls to top of deck using lag screws item 36 on 12-inch centers. Fasten walls together using 3-inch deck screws item 39 on 6-inch centers. Install remaining roof trusses item 24 on 16-inch centers. Attach plywood roof deck item 20 to trusses using 1 ⅝-inch deck screws item 38 on 6-inch centers. Install shingles per manufacturer's recommended nailing pattern. Decorate to suit with windows, table, bunk beds, and flower box.

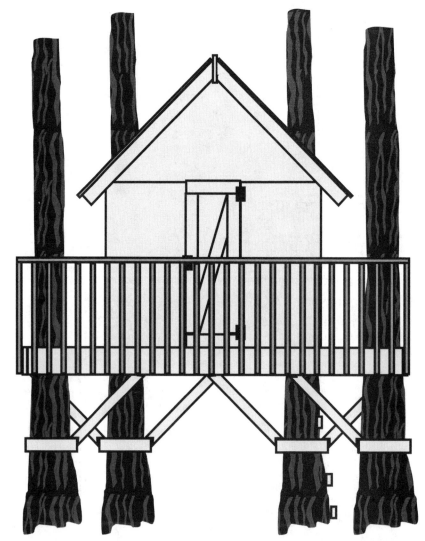

Double-Deck Cabin Tree House Plan

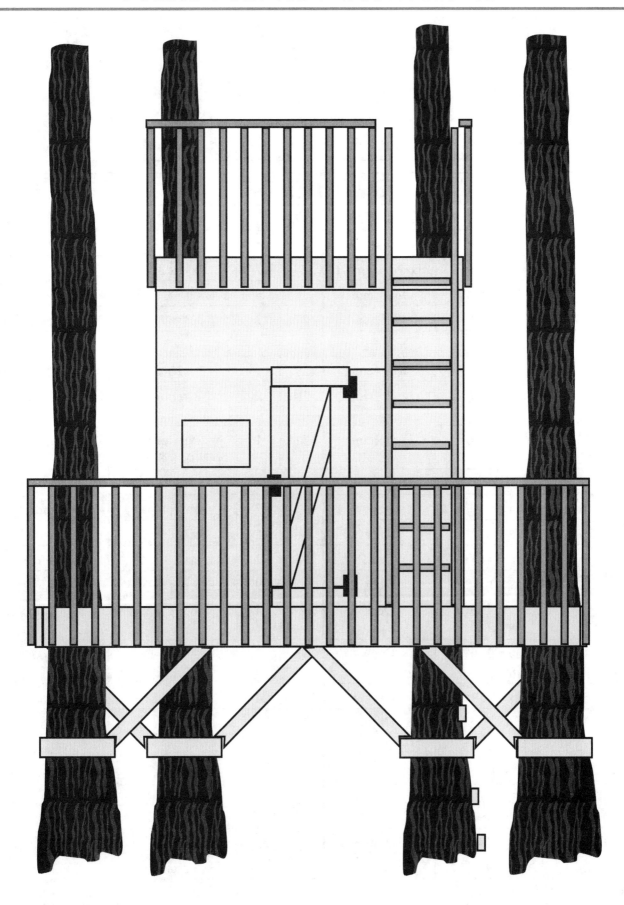

Double-Deck Cabin Tree House Plan

TOOLS REQUIRED

- Safety glasses
- Gloves
- Carpenter's hammer
- 24" framing square
- 24" level
- Hand drill
- Screwdriver
- Tape measure
- Hand saw
- $\frac{3}{16}$" drill bit
- Paint brush
- $\frac{3}{4}$" wrench

BILL OF MATERIAL

ITEM	QTY.	SIZE	DESCRIPTION	ITEM	QTY.	SIZE	DESCRIPTION
1	8	2 x 6 x 2'-0"	Lower support	24	4	2 x 4 x 3'-0"	Wall frame
2	8	2 x 6 x 2'-3"	Lower support	25	2	2 x 4 x 2'-0"	Wall frame
3	7	2 x 12 x 14'-0"	Deck frame	26	8	2 x 10 x 7'-9"	Upper deck frame
4	9	2 x 12 x 1'-10 ½"	Deck frame	27	2	2 x 10 x 10'-0"	Upper deck frame
5	13	2 x 12 x 2'-0"	Deck frame	28	7	2 x 10 x 1'-2 ½"	Upper deck frame
6	9	2 x 12 x 3'-0"	Deck frame	29	1	2 x 10 x 5"	Upper deck frame
7	1	2 x 12 x 3'-6"	Deck frame	30	2	¾" exterior ply x 4'-0" x 8'-0"	Upper deck
8	7	2 x 12 x 7'-6"	Deck frame	31	1	¾" exterior ply x 2'-0" x 8'-0"	Upper deck
9	6	2 x 12 x 1'-2 ¾"	Deck frame	32	4	2 x 4 x 10'-0"	Handrail
10	10	2 x 4 x 1'-6"	Steps	33	2	2 x 4 x 8'-0"	Handrail
11	20	2 x 6 x 3'-6"	Gusset	34	2	2 x 4 x 5'-9"	Handrail
12	5	⁵⁄₄ x 6 x 9'-10 ½"	Deck	35	2	2 x 4 x 12'-0"	Ladder
13	3	⁵⁄₄ x 6 x 14'-2"	Deck	36	9	2 x 4 x 1'-6"	Ladder
14	2	½" ply x 4'-0" x 8'-0"	Deck	37	4	⁵⁄₄ x 6 x 1'-11 ½"	Door
15	1	½" ply x 2'-0" x 8'-0"	Deck	38	2	½" ply x 1'-11 ½" x 5'-11 ½"	Door
16	5	⁵⁄₄ x 6 x 7'-10 ½"	Deck	39	4	⁵⁄₄ x 6 x 4'-11 ½"	Door
17	3	⁵⁄₄ x 6 x 9'-8"	Deck	40	2	⁵⁄₄ x 6 x 5'-2"	Door
18	130	2 x 2 x 4'-0"	Post	41	4	2"	Hinge
19	8	2 x 4 x 14'-0"	Handrail	42	2	Standard door size	Latch
20	13	½" exterior ply cut to suit	Deck, walls	43	box	½" x 4" lag screw	Lag screws
21	16	2 x 4 x 7'-9"	Wall frame	44	box	1" deck screws	Deck screws
22	2	2 x 4 x 9'-5"	Wall base	45	box	1 ⁵⁄₈" deck screws	Deck screws
23	2	2 x 4 x 8'-0"	Wall frame	46	box	3" deck screws	Deck screws

Double-Deck Cabin Tree House Plan

STEP 1 – ASSEMBLE LOWER-SUPPORT FRAMES:

The best trees for the construction of a tree house are apple, ash, beech, cedar, Colorado blue spruce, cypress, fir, hemlock, hickory, sugar maple, oak, and weeping willow. Attach items 1 & 2, qty. eight each to the tree trunk using item 43 lag screws, qty. four per board. Use a ¼-inch drill bit and drill pilot holes to prevent wood from splitting. The elevation of the support should be 2 feet below the intended height of the tree house's main upper deck. As you install each board, install one screw and then check for level prior to installing the subsequent screws.

STEP 2 – ASSEMBLE MAIN SUPPORT FRAME:

Assemble the main frame by first attaching item 3 to the tree trunks at the desired elevation. Fasten securely using item 43 lag screws, qty. eight per connection. Drill a pilot hole using a ¼-inch drill bit to prevent the wood from splitting. As you begin, confirm the first beam item 3 is level. All subsequent beams must be installed at the same elevation. Continue to assemble the remaining items 3, 4, 5, 6, 7, 8, & 9 using lag screws item 43, qty. four per connection.

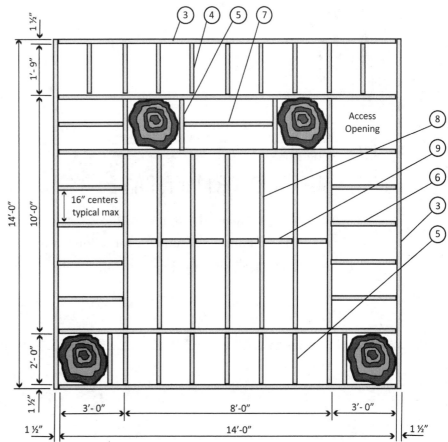

Locate a cluster of 4 trees aligned similarly to this illustration. One large tree with multiple branches in similar locations will also suffice. Adjust the floor plan as required to suit your specific circumstances.

Double-Deck Cabin Tree House Plan

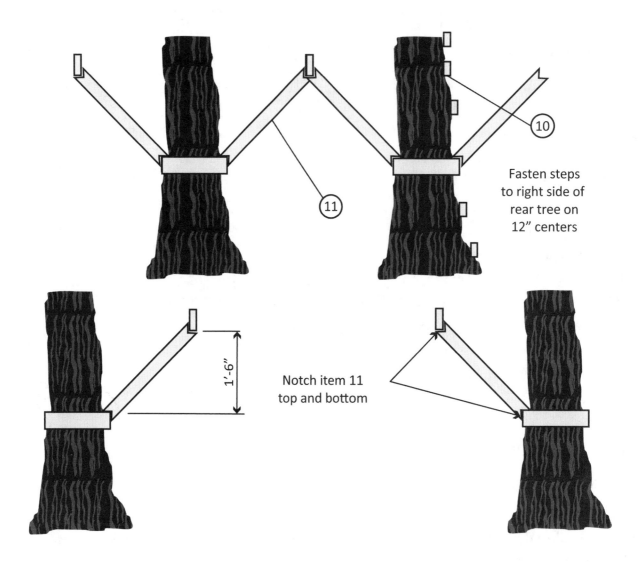

STEP 3 – INSTALL
GUSSET SUPPORTS FOR MAIN FRAME:

Notch item 11 gusset at top and bottom to match lower support and upper main deck. Fasten to frame using item 43 lag bolts, qty. four per connection. Fasten steps item 10 to right side of right rear tree using item 43 lag bolts, qty. four per connection.

Double-Deck Cabin Tree House Plan

STEP 4 – INSTALL MAIN DECK:

Assemble the main deck. Attach plywood deck to frame using 1 ⅝-inch deck screws item 45 on 6-inch spacing. Attach perimeter deck to frame using 1 ⅝-inch deck screws item 45, qty. three per connection. Install railing supports item 18 around perimeter of deck using 3-inch deck screws item 46, qty. four per connection. Maximum spacing of item 18 is 4 inches. Fasten railing item 19 to top and inside of posts using 3-inch deck screws item 46, qty. two per connection.

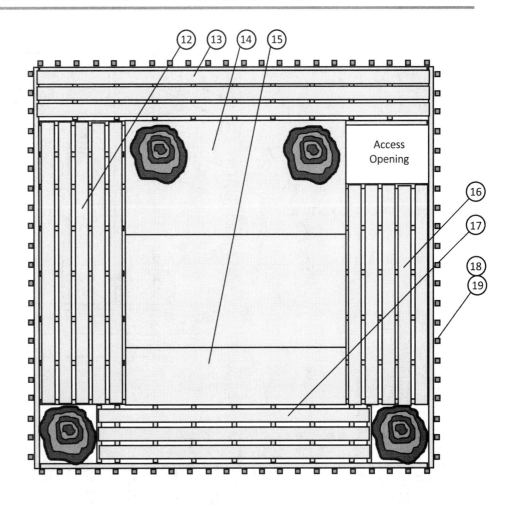

STEP 5 – ASSEMBLE SIDE WALLS:

Splice plywood item 20 to match overall dimensions as shown. Attach plywood to frame items 21 & 22 using 1 ⅝-inch deck screws item 45 on 6-inch spacing.

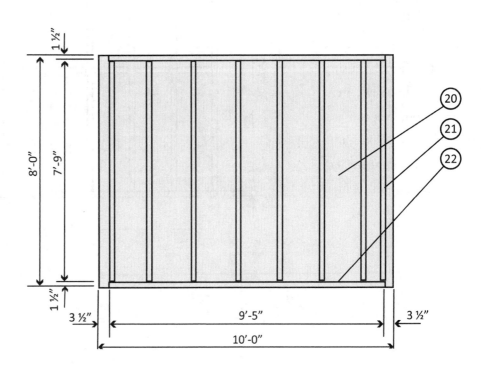

Double-Deck Cabin Tree House Plan

STEP 6 – ASSEMBLE END WALLS:

Cut out plywood end walls item 20 to match overall dimensions as shown. Assemble all four walls on top of plywood deck. Attach walls to deck using item 43 lag screws on 12-inch spacing.

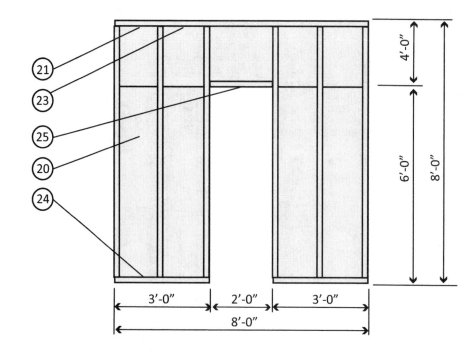

STEP 7 – ASSEMBLE UPPER-DECK SUPPORT FRAME:

Assemble upper deck support frame using 3-inch deck screws item 46, qty. four per connection. Attach deck frame to top of walls using lag screws item 43 on 12-inch centers.

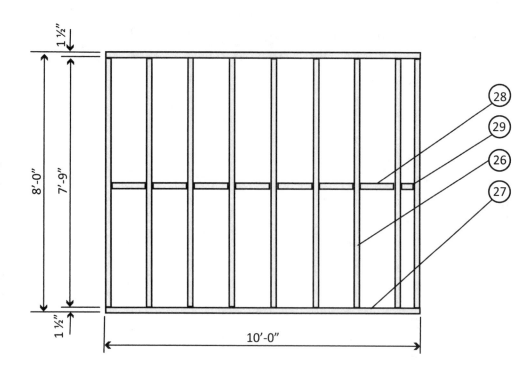

Double-Deck Cabin Tree House Plan

STEP 8 – INSTALL PLYWOOD DECK:

Cutout plywood deck item 20 to match overall dimensions as shown. Attach to deck support frame using 1 ⅝-inch deck screws item 45 on 6-inch centers. **NOTE:** Wood deck must be exterior-grade, pressure-treated plywood.

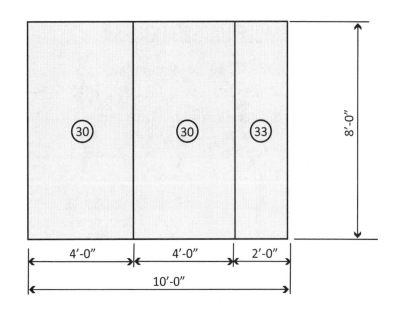

STEP 9 – INSTALL HANDRAIL AROUND UPPER DECK:

Install railing supports item 18 around perimeter of deck using 3-inch deck screws item 46, qty. four per connection. Maximum spacing of item 18 is 4 inches. Fasten railing item 32, 33, & 34 to top and inside of posts using 3-inch deck screws item 46, qty. two per connection.

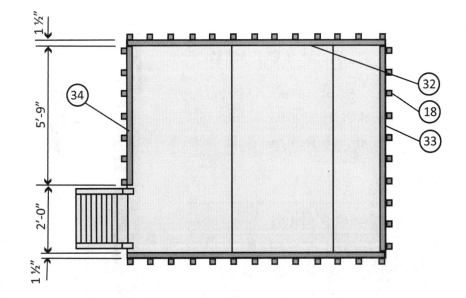

STEP 10 – ASSEMBLE ACCESS LADDER:

Assemble access ladder items 35 & 36 using 3-inch deck screws item 46, qty. four per connection. Attach ladder to upper deck using lag screws item 43, qty. two per connection.

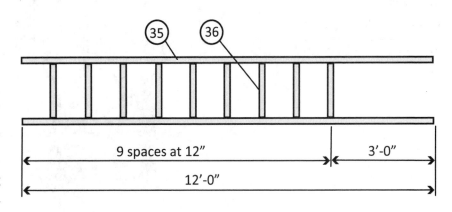

Double-Deck Cabin Tree House Plan

STEP 11 – ASSEMBLE HINGED DOORS:

Attach door frame items 37, 30, & 40 to plywood door item 38 using 1-inch deck screws item 44 on 6-inch centers. Install hinges item 42 and latch item 41.

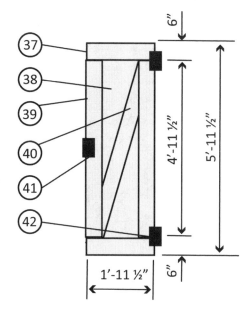

STEP 12 – FINAL ASSEMBLY:

Paint and decorate to suit. Add windows, table, bunk beds, and flower box. Build portable furniture for upper deck.

> ### CAUTION
>
> *Wear protective safety glasses when using hammer or operating power tools, and wear protective gloves when handling sharp objects and splintered wood.*

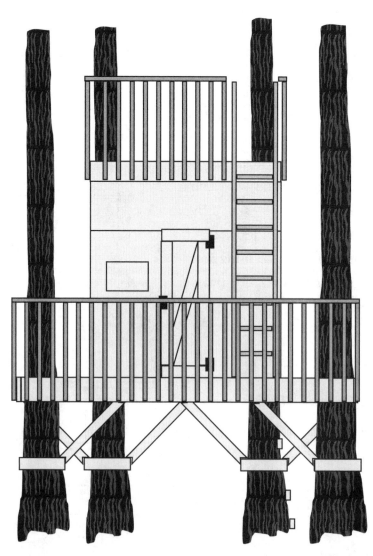

A-Frame Cabin Tree House With Loft Plan

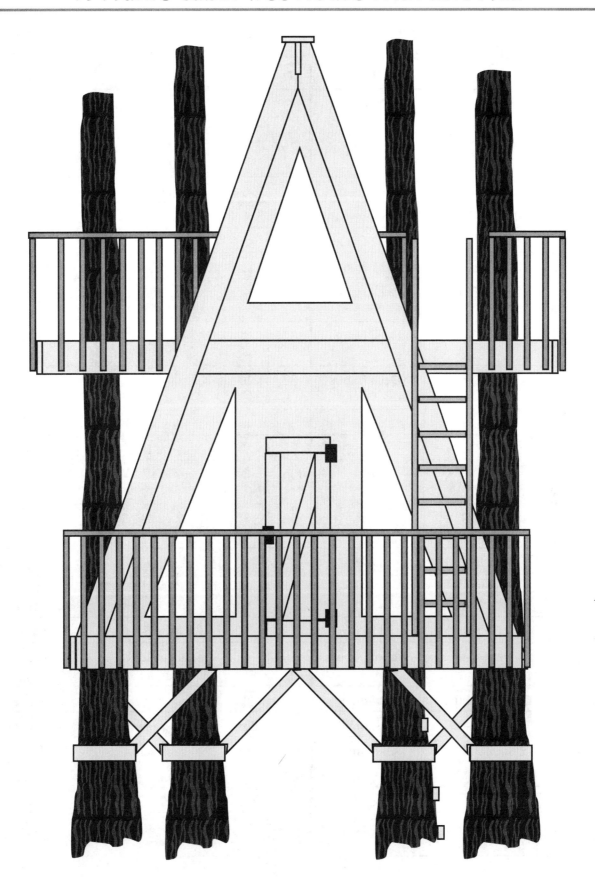

A-Frame Cabin Tree House With Loft Plan

BILL OF MATERIAL							
ITEM	QTY.	SIZE	DESCRIPTION	ITEM	QTY.	SIZE	DESCRIPTION
1	8	2 x 6 x 2'-0"	Lower support	24	2	¾ ply x 4'-0" x 7'-4"	Upper deck
2	8	2 x 6 x 2'-3"	Lower support	25	1	¾ ply x 2'-0" x 7'-4"	Upper deck
3	7	2 x 12 x 14'-0"	Deck frame	26	6	2 x 4 x 5'-6"	Handrail
4	9	2 x 12 x 1'-10 ½"	Deck frame	27	4	2 x 4 x 10'-0"	Handrail
5	13	2 x 12 x 2'-0"	Deck frame	28	2	2 x 4 x 2'-0"	Handrail
6	10	2 x 12 x 3'-0"	Deck frame	29	2	2 x 4 x 1'-6"	Handrail
7	1	2 x 12 x 3'-6"	Deck frame	30	4	½" ply x 2'-7" x 7'-0"	Door frame
8	7	2 x 12 x 7'-6"	Deck frame	31	2	½" ply x 2'-6" x 2'-7"	Door frame
9	6	2 x 12 x 1'-2 ¾"	Deck frame	32	20	½" ply x 4'-0" x 8'-0" cut to suit	Walls
10	10	2 x 4 x 1'-6"	Steps	33	20	2 x 4 x 10'-0"	Wall studs
11	20	2 x 6 x 3'-6"	Gusset	34	9	2 x 4 x 1'-6"	Ladder
12	3	¾ ply x 4'-0" x 6'-0"	Deck	35	2	2 x 4 x 12'-0"	Ladder
13	3	¾ ply x 4'-0" x 8'-0"	Deck	36	8	5/4 x 6 x 1'-11 ½"	Door
14	3	5/4 x 6 x 8'-11"	Deck	37	4	½" ply x 1'-11 ½" x 5'-11 ½"	Door
15	3	5/4 x 6 x 2'-11"	Deck	38	8	5/4 x 6 x 4'-11 ½"	Door
16	3	5/4 x 6 x 9'-8"	Deck	39	4	5/4 x 6 x 5'-2"	Door
17	96	2 x 2 x 4'-0"	Handrail post	40	4	Standard door latch	Latch
18	4	2 x 4 x 14'-0"	Handrail	41	8	2"	Hinge
19	8	2 x 4 x 2'-0"	Handrail	42	box	½" x 4" lag screw	Lag screws
20	18	2 x 8 x 21'-2 ¼"	A-Frame	43	box	1" deck screws	Deck screws
21	2	2x 12 x 12'-0"	A-Frame	44	box	1 5/8" deck screws	Deck screws
22	2	2 x 12 x 10'-0"	Upper deck	45	box	3" deck screws	Deck screws
23	9	2 x 12 x 16'-0"	Upper deck	46	2 gal.	Optional	Paint

A-Frame Cabin Tree House With Loft Plan

TOOLS REQUIRED	
• Safety glasses	• Screwdriver
• Gloves	• Tape measure
• Carpenter's hammer	• Hand saw
• 24" framing square	• ³⁄₁₆" drill bit
• 24" level	• Paint brush
• Hand drill	• ¾" wrench

STEP 1 – ASSEMBLE LOWER SUPPORT FRAMES:

The best trees for the construction of a tree house are apple, ash, beech, cedar, Colorado blue spruce, cypress, fir, hemlock, hickory, sugar maple, oak, and weeping willow. Attach items 1 & 2, qty. eight each to the tree trunk using item 42 lag screws qty. four per board. Use a ¼-inch drill bit and drill pilot holes to prevent wood from splitting. The elevation of the support should be 2 feet below the intended height of the tree house's main upper deck. As you install each board, install one screw and then check for level prior to installing the subsequent screws.

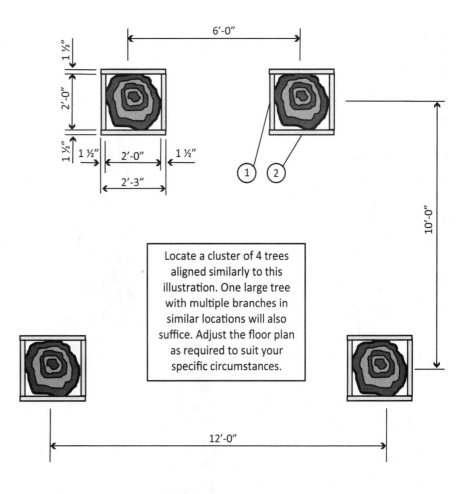

Locate a cluster of 4 trees aligned similarly to this illustration. One large tree with multiple branches in similar locations will also suffice. Adjust the floor plan as required to suit your specific circumstances.

A-Frame Cabin Tree House With Loft Plan

STEP 2 – ASSEMBLE MAIN SUPPORT FRAME:

Assemble the main frame by first attaching item 3 to the tree trunks at the desired elevation. Fasten securely using item 42 lag screws, qty. eight per connection. Drill a pilot hole using a ¼-inch drill bit to prevent the wood from splitting. As you begin, confirm the first beam item 3 is level. All subsequent beams must be installed at the same elevation. Continue to assemble the remaining items 3, 4, 5, 6, 7, 8, & 9 using lag screws item 42, qty. four per connection.

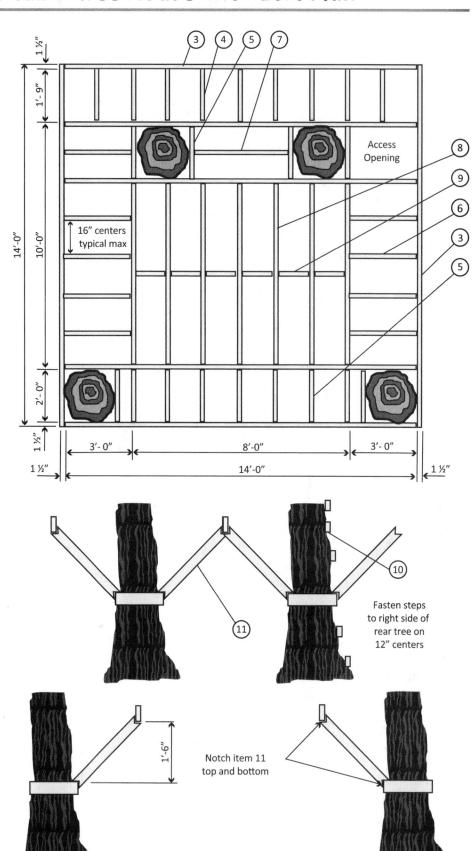

STEP 3 – INSTALL GUSSET SUPPORTS FOR MAIN FRAME:

Notch item 11 gusset at top and bottom to match lower support and upper main deck. Fasten to frame using item 42 lag bolts, qty. four per connection. Fasten steps item 10 to right side of right rear tree using item 42 lag bolts, qty. four per connection.

A-Frame Cabin Tree House With Loft Plan

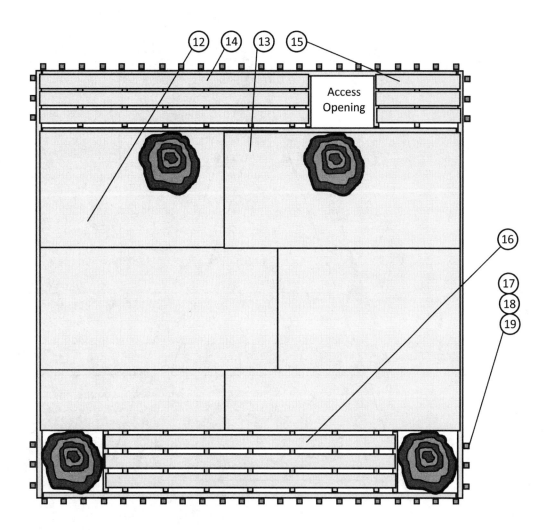

STEP 4 – INSTALL MAIN DECK:

Assemble the main deck. Attach plywood deck to frame using 1 $\frac{5}{8}$-inch deck screws item 44 on 6-inch spacing. Attach perimeter deck to frame using 1 $\frac{5}{8}$-inch deck screws item 44, qty. three per connection. Install railing supports item 17 around perimeter of deck using 3-inch deck screws item 45, qty. four per connection. Maximum spacing of item 17 is 4 inches. Fasten railing item 19 to top and inside of posts using 3-inch deck screws item 45, qty. two per connection.

A-Frame Cabin Tree House With Loft Plan

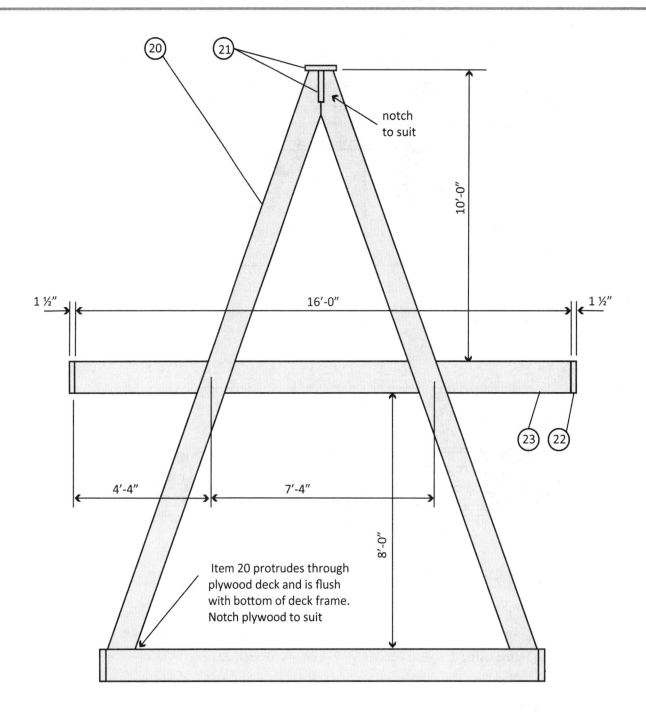

notch
to suit

10'-0"

1 ½"   16'-0"   1 ½"

4'-4"   7'-4"

8'-0"

Item 20 protrudes through
plywood deck and is flush
with bottom of deck frame.
Notch plywood to suit

STEP 5 – ASSEMBLE A-FRAME STRUCTURE:

Notch plywood deck to accommodate item 20 which mounts flush with the bottom of deck frame. Attach item 20 to lower deck using lag screws item 42 qty. eight screws per connection. Notch top of item 20 to mate accordingly. Item 21 extends 12 inches beyond A-Frame. Attach upper deck item 23 to item 20 using lag screws item 42, qty. eight per connection.

A-Frame Cabin Tree House With Loft Plan

STEP 6 – ASSEMBLE UPPER-DECK SUPPORT FRAME:

Assemble upper deck support frame using 3-inch deck screws item 45, qty. six per connection. Attach upper deck to A-Frame item 20 using lag screws item 42, qty. eight per connection.

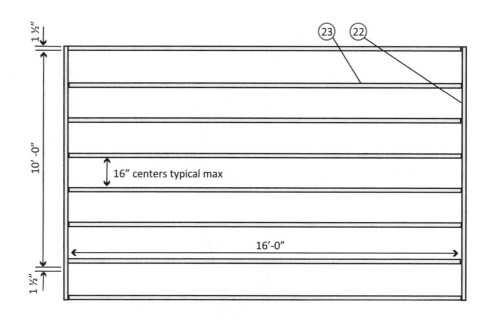

STEP 7 – INSTALL UPPER PLYWOOD DECK:

Attach plywood deck to support frame using 1 ⅝-inch deck screws item 44 on 6-inch centers. Install perimeter handrail and supports using 3-inch deck screws item 45.

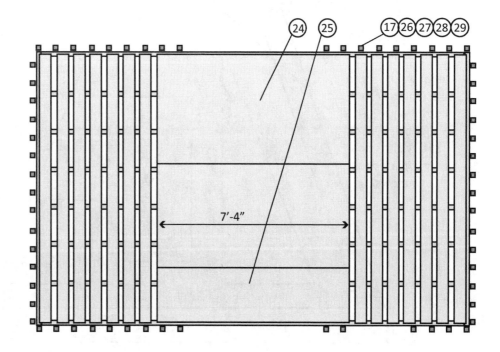

A-Frame Cabin Tree House With Loft Plan

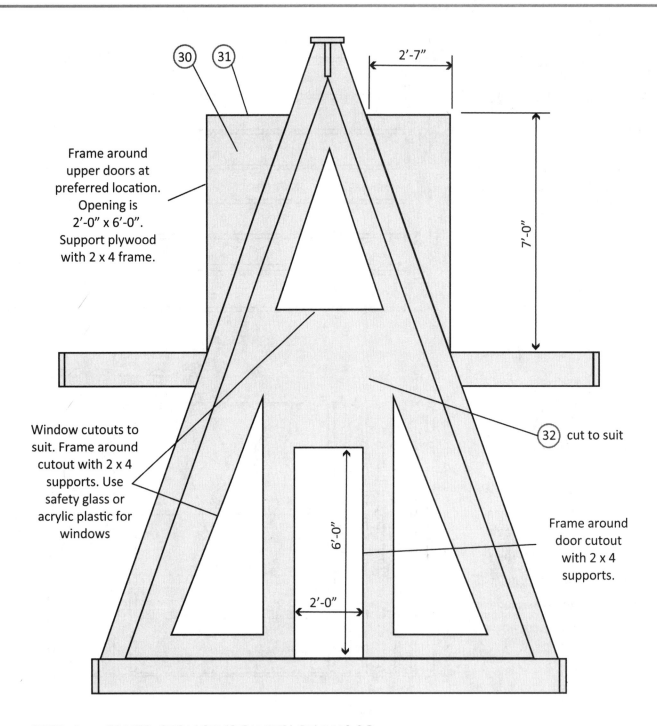

Frame around upper doors at preferred location. Opening is 2'-0" x 6'-0". Support plywood with 2 x 4 frame.

Window cutouts to suit. Frame around cutout with 2 x 4 supports. Use safety glass or acrylic plastic for windows

cut to suit

Frame around door cutout with 2 x 4 supports.

STEP 8 – COVER STRUCTURE WITH PLYWOOD:

Cover entire A-Frame with ½-inch plywood item 32, cut to suit. Attach item 32 to structure using 1 ⅝-inch deck screws item 44 on 6-inch spacing. Frame around door and window openings using 2x4s item 33 to support plywood. Cut to suit as required. Assemble item 33 using 3-inch deck screws item 45, qty. two per connection.

A-Frame Cabin Tree House With Loft Plan

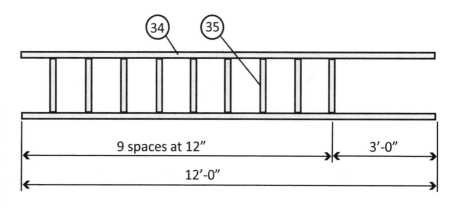

STEP 9 – ASSEMBLE ACCESS LADDER:

Assemble access ladder items 34 & 35 using 3-inch deck screws item 45, qty. four per connection. Attach ladder to upper deck using lag screws item 42, qty. two per connection.

STEP 10 – ASSEMBLE HINGED DOORS, QTY. FOUR:

Attach door frame items 36, 38, & 40 to plywood door item 37 using 1-inch deck screws item 43 on 6-inch centers. Install hinges item 41 and latch item 40.

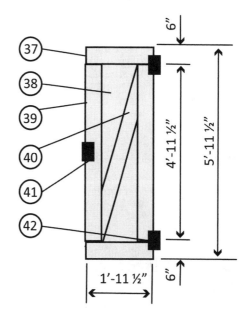

STEP 11 – FINAL ASSEMBLY:

Paint and decorate to suit. Add windows, table, bunk beds, and flower box. Build portable furniture for upper deck.

CAUTION

Wear protective safety glasses when using hammer or operating power tools, and wear protective gloves when handling sharp objects and splintered wood.

Two-Story Tree House Plan

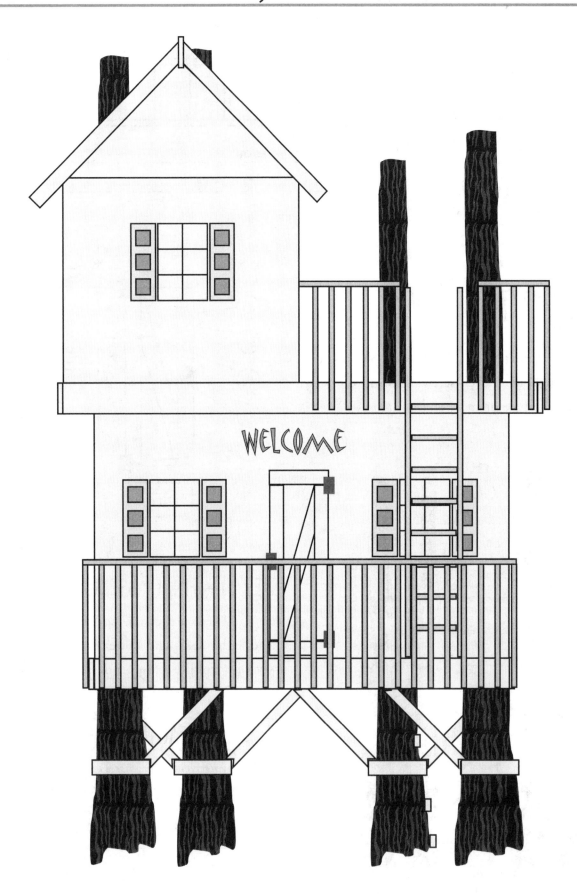

WELCOME

Two-Story Tree House Plan

BILL OF MATERIAL							
ITEM	QTY.	SIZE	DESCRIPTION	ITEM	QTY.	SIZE	DESCRIPTION
1	8	2 x 6 x 2'-0"	Lower support	31	2	¾ ply x 2'-0" x 8'-0"	Deck
2	8	2 x 6 x 2'-3"	Lower support	32	2	2 x 4 x 8'-0"	Handrail
3	7	2 x 12 x 14'-0"	Deck frame	33	2	2 x 4 x 10'-0"	Handrail
4	9	2 x 12 x 1'-10 ½"	Deck frame	34	2	2 x 4 x 2'-6"	Handrail
5	13	2 x 12 x 2'-0"	Deck frame	35	2	2 x 4 x 3'-6"	Handrail
6	10	2 x 12 x 3'-0"	Deck frame	36	13	¾ ply x 4'-0" x 8'-0"	Deck
7	1	2 x 12 x 3'-6"	Deck frame	37	16	2 x 4 x 6'-3"	Wall studs
8	7	2 x 12 x 7'-6"	Deck frame	38	4	2 x 4 x 9'-5"	Wall studs
9	6	2 x 12 x 1'-2 ¾"	Deck frame	39	2	2 x 4 x 3'-8 ½"	Wall studs
10	10	2 x 4 x 1'-6"	Steps	40	1	2 x 12 x 12'-0"	Roof truss
11	20	2 x 6 x 3'-6"	Gusset	41	16	2 x 8 x 7'-0"	Roof truss
12	3	¾ ply x 4'-0" x 6'-0"	Deck	42	4	2 x 4 x 8'-10 ½"	Wall studs
13	3	¾ ply x 4'-0" x 8'-0"	Deck	43	4	2 x 4 x 7'-7 ½"	Wall studs
14	3	5/4 x 6 x 8'-11"	Deck	44	4	2 x 4 x 6'-4 ½"	Wall studs
15	3	5/4 x 6 x 2'-11"	Deck	45	2	2 x 4 x 8'-0"	Wall studs
16	3	5/4 x 6 x 9'-8"	Deck	46	bdl.	Roof standard	Shingles
17	96	2 x 2 x 4'-0"	Handrail post	47	2	2 x 4 x 12'-0"	Ladder
18	4	2 x 4 x 14'-0"	Handrail	48	9	2 x 4 x 1'-6"	Ladder
19	8	2 x 4 x 2'-0"	Handrail	49	6	5/4 x 6 x 1'-11 ½"	Door
20	8	2 x 4 x 1'-9"	Wall studs	50	3	½" ply x 1'-11 ½" x 5'-11 ½"	Door
21	18	2 x 4 x 2'-0"	Wall studs	51	6	5/4 x 6 x 4'-11 ½"	Door
22	32	2 x 4 x 7'-9"	Wall studs	52	3	5/4 x 6 x 5'-2"	Door
23	21	½ ply x 4'-0" x 8'-0"	Walls/roof	53	3	Standard door size	Latch
24	6	2 x 4 x 2'-9"	Wall studs	54	6	2"	Hinge
25	4	2 x 4 x 9'-5"	Wall studs	55	box	½" x 4" lag screws	Screws
26	4	2 x 4 x 6'-0"	Wall studs	56	box	1" deck screws	Screws
27	8	2 x 12 x 16'-0"	Deck	57	box	1 5/8" deck screws	Screws
28	2	2 x 12 x 10'-0"	Deck	58	box	3" deck screws	Screws
29	14	2 x 12 x 1'-2 ½"	Deck	59	box	1" roofing nails	Nails
30	2	¾" ply x 2'-0" x 6'-0"	Deck	60	2 gal.	Optional	Paint

Two-Story Tree House Plan

TOOLS REQUIRED

- Safety glasses
- Gloves
- Carpenter's hammer
- 24" framing square
- 24" level
- Hand drill
- Screwdriver
- Tape measure
- Hand saw
- $\frac{3}{16}$" drill bit
- Paint brush
- ¾" wrench

STEP 1 – ASSEMBLE LOWER SUPPORT FRAMES:

The best trees for the construction of a tree house are apple, ash, beech, cedar, Colorado blue spruce, cypress, fir, hemlock, hickory, sugar maple, oak, and weeping willow. Attach items 1 & 2, qty. eight each to the tree trunk using item 55 lag screws qty. four per board. Use a ¼-inch drill bit and drill pilot holes to prevent wood from splitting. The elevation of the support should be 2 feet below the intended height of the tree house's main upper deck. As you install each board, install one screw and then check for level prior to installing the subsequent screws.

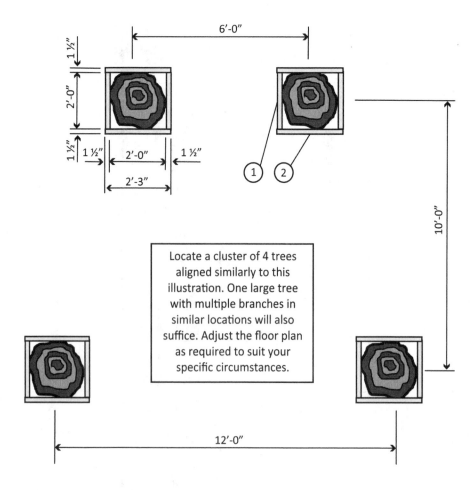

Locate a cluster of 4 trees aligned similarly to this illustration. One large tree with multiple branches in similar locations will also suffice. Adjust the floor plan as required to suit your specific circumstances.

Two-Story Tree House Plan

STEP 2 – ASSEMBLE MAIN SUPPORT FRAME:

Assemble the main frame by first attaching item 3 to the tree trunks at the desired elevation. Fasten securely using item 55 lag screws, qty. eight per connection. Drill a pilot hole using a ¼-inch drill bit to prevent the wood from splitting. As you begin, confirm the first beam item 3 is level. All subsequent beams must be installed at the same elevation. Continue to assemble the remaining items 3, 4, 5, 6, 7, 8, & 9 using lag screws item 55, qty. four per connection.

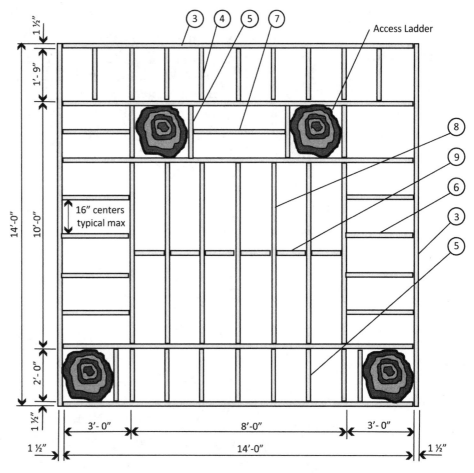

STEP 3 – INSTALL GUSSET SUPPORTS FOR MAIN FRAME:

Notch item 11 gusset at top and bottom to match lower support and upper main deck. Fasten to frame using item 55 lag bolts, qty. four per connection. Fasten steps item 10 to right side of right rear tree using item 55 lag bolts, qty. four per connection.

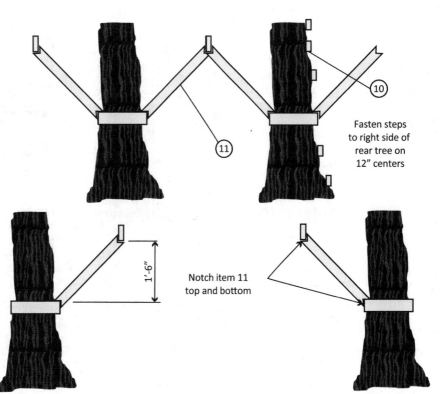

Two-Story Tree House Plan

STEP 4 – INSTALL MAIN DECK:

Assemble the main deck. Attach plywood deck to frame using 1 ⅝-inch deck screws item 57 on 6-inch spacing. Attach perimeter deck to frame using 1 ⅝-inch deck screws item 57, qty. three per connection. Install railing supports item 17 around perimeter of deck using 3-inch deck screws item 58, qty. four per connection. Maximum spacing of item 17 is 4 inches. Fasten railing item 19 to top and inside of posts using 3-inch deck screws item 58, qty. two per connection.

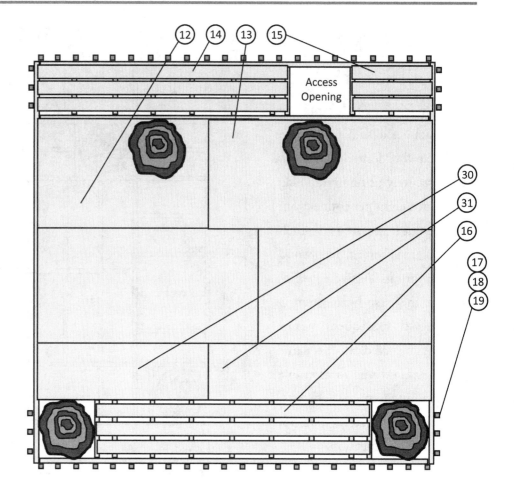

STEP 5 – ASSEMBLE 1ST-FLOOR SIDE WALLS, QTY. TWO:

Assemble wall frame items 20, 21, 22, 24, & 25 using 3-inch deck screws item 58, qty. two per connection. Cut plywood item 23 to suit and attach to frame using 1 ⅝-inch deck screws item 58 on 6-inch spacing. Attach wall assembly to deck using item 55 lag screws on 12-inch spacing.

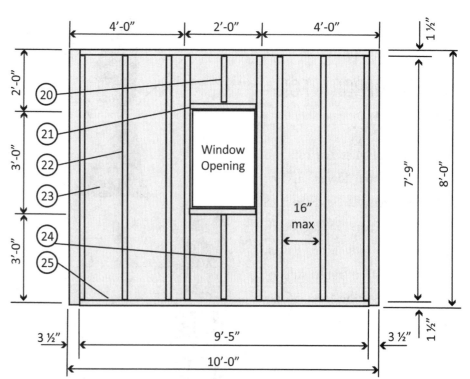

Two-Story Tree House Plan

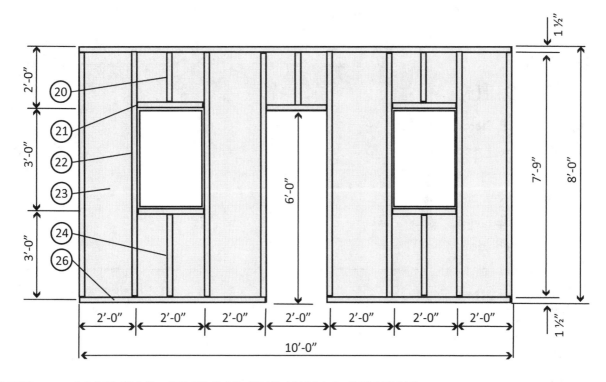

STEP 6 – ASSEMBLE 1ST-FLOOR END WALLS, QTY. TWO:

Assemble wall frame items 20, 21, 22, 24, & 26 using 3-inch deck screws item 58, qty. two per connection. Cut plywood item 23 to suit and attach to frame using 1 ⅝-inch deck screws item 57 on 6-inch spacing. Attach wall assembly to deck using item 55 lag screws on 12-inch spacing.

STEP 7 – ASSEMBLE UPPER-DECK SUPPORT FRAME:

Assemble upper deck support frame items 27, 28, & 29 using 3-inch deck screws item 58, qty. six per connection. Attach upper deck to A-Frame item 20 using lag screws item 55, qty. eight per connection.

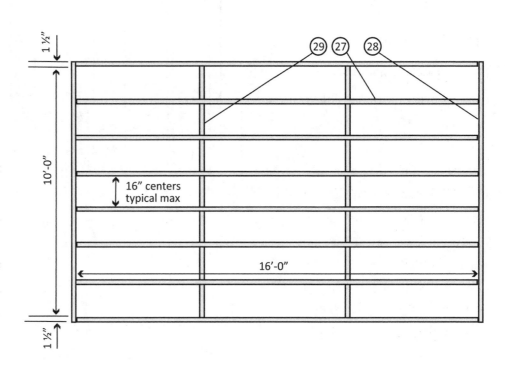

Two-Story Tree House Plan

STEP 8 – INSTALL UPPER PLYWOOD DECK:

Attach plywood deck items 13 & 31 to support frame using 1 ⅝-inch deck screws item 57 on 6-inch centers. Install perimeter handrail and supports using 3-inch deck screws item 58. Install item 36 deck boards using 1 ⅝-inch deck screws item 57, qty. three per connection.

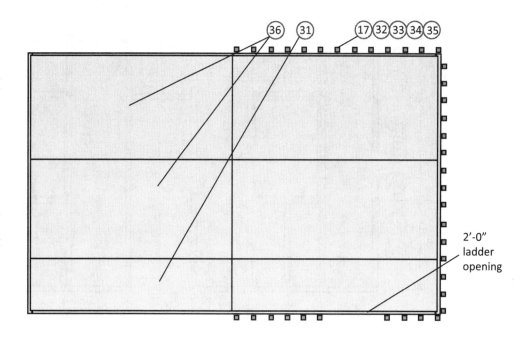

STEP 9 – ASSEMBLE UPPER LEFT-SIDE WALL:

Splice plywood item 23 to match overall dimensions as shown. Attach plywood to frame items 37, 38, & 39 using 1 ⅝-inch deck screws on 6-inch spacing. Attach assembly to deck using item 55 lag screws on 12-inch spacing.

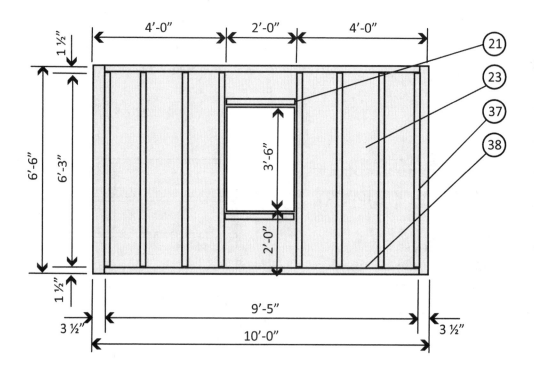

Two-Story Tree House Plan

STEP 10 – ASSEMBLE UPPER RIGHT-SIDE WALL:

Splice plywood item 23 to match overall dimensions as shown. Attach plywood to frame items 21, 37, & 39 using 1 ⅝-inch deck screws on 6-inch spacing. Attach assembly to deck using item 55 lag screws on 12-inch spacing.

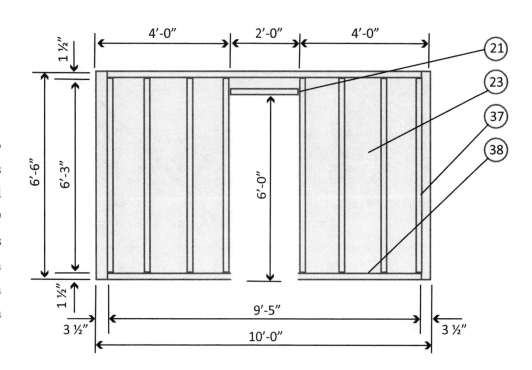

STEP 11 – ASSEMBLE UPPER END WALLS:

Assemble plywood end wall item 23 and end wall frame items 21, 41, 42, 43, 44, & 45 using 1 ⅝-inch deck screws item 57 on 6-inch spacing. Attach assembly to top side of deck using lag screws item 55 on 12-inch spacing.

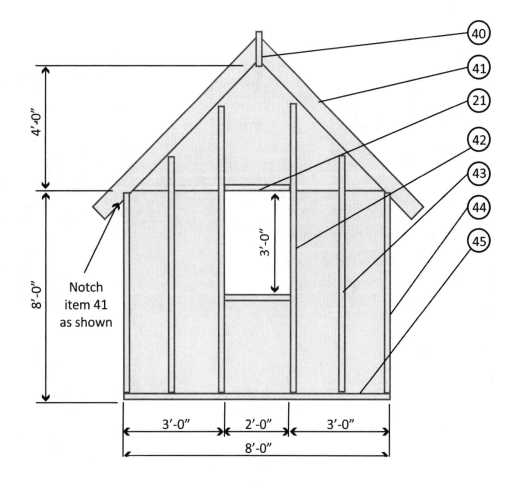

Notch item 41 as shown

Two-Story Tree House Plan

STEP 12 – INSTALL ROOF DECK & SHINGLES:

Attach roof deck item 23 using 1-inch deck screws item 56 on 6-inch spacing. Install shingles item 46 using manufacturer's recommended nailing pattern.

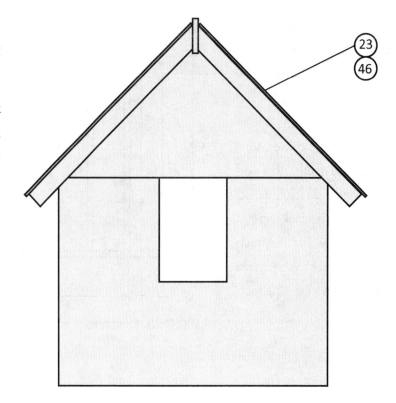

STEP 13 – ASSEMBLE ACCESS LADDER:

Assemble access ladder items 47 & 48 using 3-inch deck screws item 58, qty. four per connection. Attach ladder to upper deck using lag screws item 55, qty. two per connection.

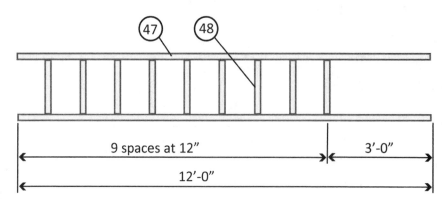

Two-Story Tree House Plan

STEP 14 – ASSEMBLE HINGED DOORS, QTY. THREE:

Attach door frame items 49, 51, & 52 to plywood door item 50 using 1-inch deck screws item 56 on 6-inch centers. Install hinges item 54 and latch item 53.

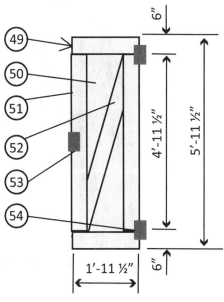

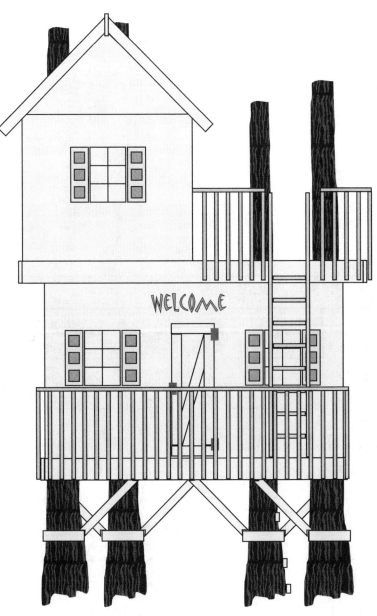

STEP 15 – FINAL ASSEMBLY:

Paint and decorate to suit. Add windows, table, bunk beds, and flower box. Build portable furniture for upper deck.

Rocket Tree House Plan

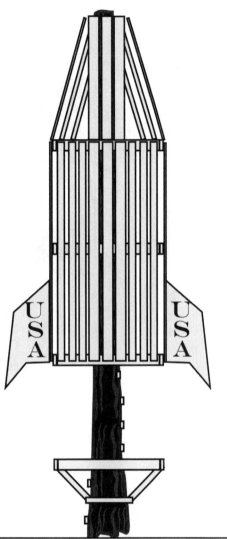

TOOLS REQUIRED

- Safety glasses
- 24" level
- Hand saw
- Gloves
- Hand drill
- $\frac{3}{16}$" drill bit
- Carpenter's hammer
- Screwdriver
- Paint brush
- 24" framing square
- Tape measure
- $\frac{9}{16}$" wrench

BILL OF MATERIAL

ITEM	QTY.	SIZE	DESCRIPTION	ITEM	QTY.	SIZE	DESCRIPTION
1	2	2 x 6 x 1'-9"	Lower support	10	12	2 x 4 x 2'-3 ½"	Deck frame
2	8	2 x 6 x 1-6"	Lower support	11	14	⁵⁄₄ x 6 x 10'-0"	Deck
3	10	2 x 6 x 5'-0"	Deck frame	12	24	⁵⁄₄ x 6 x 8'-0"	Wall
4	4	2 x 6 x 1'-7 ½"	Deck frame	13	24	⁵⁄₄ x 6 x 10-0"	Wall
5	2	2 x 6 x 5'-3"	Deck frame	14	4	½" ply x 4'-0" x 8'-0"	Wing
6	8	⁵⁄₄ x 6 x 5'-2"	Deck	15	box	½ x 4" lag screws	Screws
7	3	⁵⁄₄ x 6 x 1'-9 ½"	Deck	16	box	1" deck screws	Screws
8	8	2 x 6 x 3'-6"	Gusset	17	box	1 ⅝" deck screws	Screws
9	16	2 x 4 x 1'-6"	Steps				

Rocket Tree House Plan

STEP 1 – INSTALL LOWER SUPPORT:

Locate a tree with a trunk diameter of approximately 18 inches minimum. The best trees for construction of a tree house are apple, ash, beech, cedar, Colorado blue spruce, cypress, fir, hemlock, hickory, sugar maple, oak, and weeping willow. Attach items 1 & 2, qty. two each to the tree trunk, using item 15 lag screws, qty. four per board. The elevation of the support should be 2 feet below the intended height of the tree house floor. As you install each board, install one screw and then check for level prior to installing the subsequent screws.

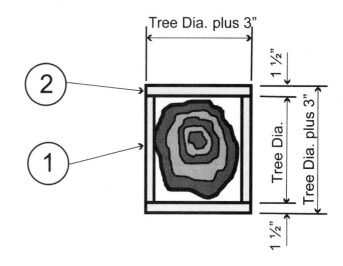

STEP 2 – INSTALL LOWER-LANDING PERIMETER FRAME:

Attach inner supports items 2 & 3 to the tree trunk, using item 15 lag screws, qty. six per board. The elevation of the support should be 2 feet above the lower support. As you install each board, install one screw, and then check for level prior to installing the subsequent screws. Install the outer supports items 3, 4, & 5 using item 15 lag screws, qty. two per connection.

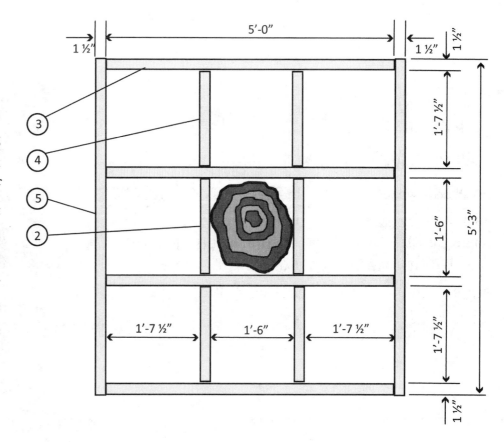

Rocket Tree House Plan

STEP 3 – INSTALL LOWER DECK:

Attach deck items 6 & 7 to the frame using item 17 deck screws 1 ⅝ inches, qty. six per connection.

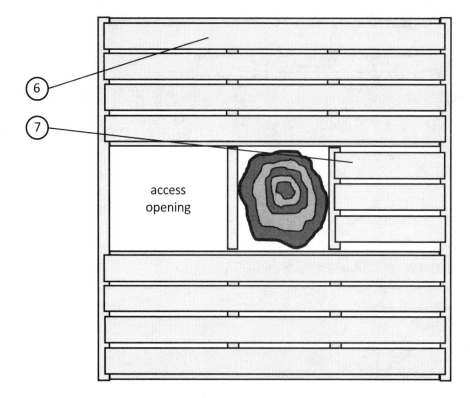

access opening

STEP 4 – INSTALL DECK SUPPORTS AND STEPS:

Install deck supports item 8 to the frame using item 15 lag screws, qty. four per connection. Attach steps item 9 to tree using lag screws item 15, qty. four per step.

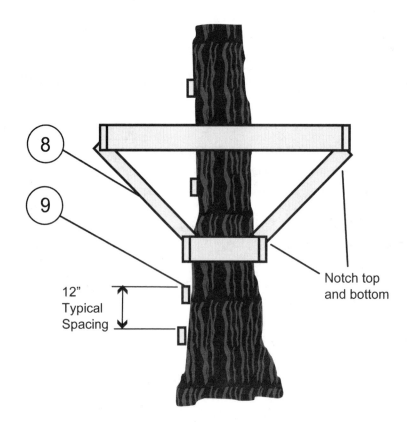

Notch top and bottom

12" Typical Spacing

Rocket Tree House Plan

STEP 5 – ASSEMBLE UPPER LANDINGS:

Attach inner supports items 2 & 3 to the tree trunk, using item 15 lag screws, qty. six per board. The elevation of the supports should be approximately 5 feet above the lower deck. As you install each board, insert one screw and then check for level prior to installing the subsequent screws. Install the outer supports items 1, 4, & 10 using item 15 lag screws, qty. two per connection.

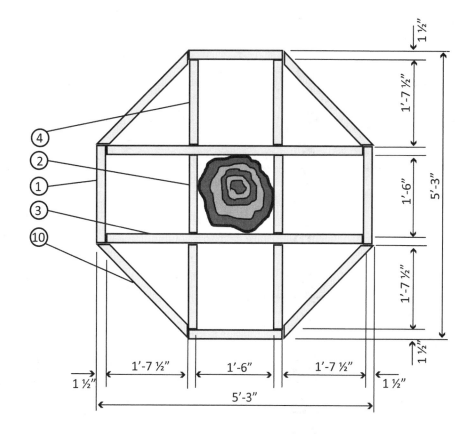

STEP 6 – INSTALL DECK ON UPPER LANDINGS:

Attach deck item 11 to the frame using item 17 deck screws 1 $\frac{5}{8}$ inches, qty. six per connection.

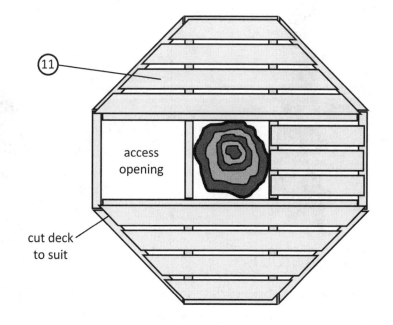

access opening

cut deck to suit

Rocket Tree House Plan

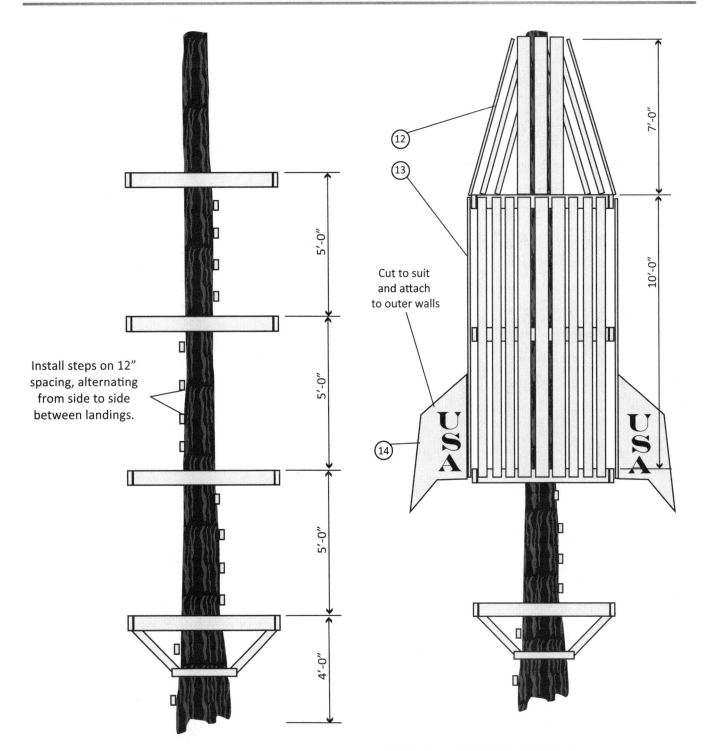

Install steps on 12" spacing, alternating from side to side between landings.

5'-0"

5'-0"

5'-0"

4'-0"

12

13

Cut to suit and attach to outer walls

14

7'-0"

10'-0"

STEP 7 – LANDING LAYOUT:

Attach landings to tree trunk, oriented as shown. Use lag screws item 15 to secure landing frame to tree. Check each landing to ensure each frame is level.

STEP 8 – INSTALL WALLS :

Attach walls to outside of frame using 1 ⅝-inch deck screws item 17, qty. three per connection. Cut item 14 to suit and attach to outer walls using 1-inch deck screws item 16 on 6-inch spacing.

Tree House Landings Plan

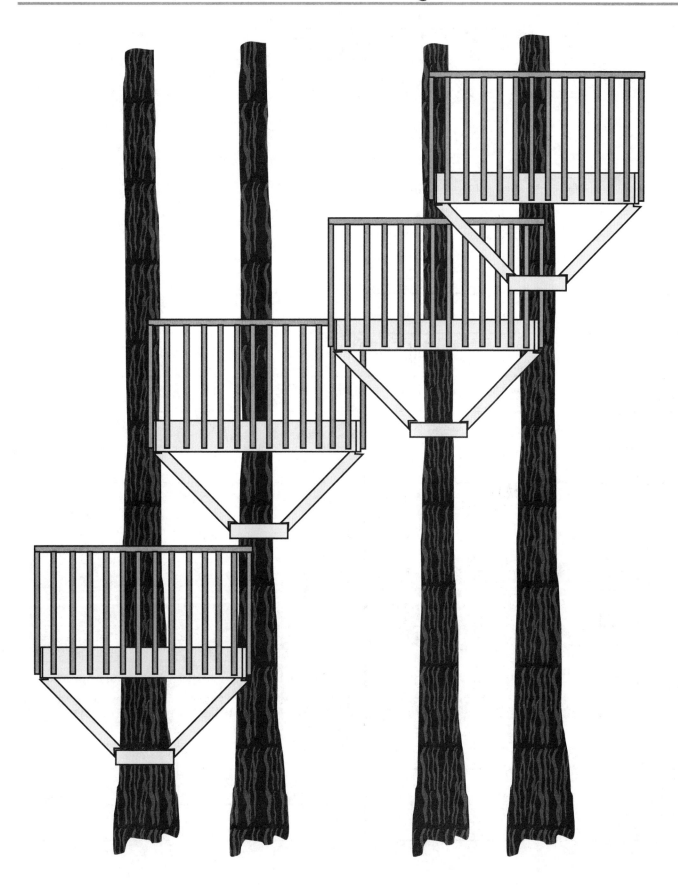

Tree House Landings Plan

TOOLS REQUIRED

- Safety glasses
- Gloves
- Carpenter's hammer
- 24" framing square
- 24" level
- Hand drill
- Screwdriver
- Tape measure
- Hand saw
- $\frac{3}{16}$" drill bit
- Paint brush
- $\frac{3}{4}$" wrench

BILL OF MATERIAL

ITEM	QTY.	SIZE	DESCRIPTION	ITEM	QTY.	SIZE	DESCRIPTION
1	8	2 x 6 x 1'-6"	Lower support	13	12	2 x 4 x 1'-7"	Handrail
2	8	2 x 6 x 1-9"	Lower support	14	12	4 x 4 x 4'-0"	Handrail
3	8	2 x 6 x 1'-6"	Deck frame	15	2	2 x 4 x 5'-2"	Handrail
4	24	2 x 6 x 5'-3"	Deck frame	16	roll	1" rope	Bridge
5	8	2 x 6 x 5'-6"	Deck frame	17	roll	½" rope	Bridge
6	32	2 x 6 x 12 ¾"	Deck frame	18	30	2 x 6 x 3'-0"	Bridge steps
7	32	2 x 6 x 3'-6"	Gusset	19	16	2 x 6 x 2'-0"	Bench
8	10	2 x 4 x 1'-6"	Steps	20	32	2 x 2 x 10"	Bench support
9	35	⁵⁄₄ x 6 x 5'-5"	Deck board	21	box	½" x 4" lag screws	Screws
10	34	⁵⁄₄ x 6 x 1'-11 ½"	Deck board	22	box	1 ⁵⁄₈" deck screws	Screws
11	8	2 x 4 x 6'-0"	Handrail	23	box	3" deck screws	Screws
12	200	2 x 2 x 4'-0"	Handrail				

Tree House Landings Plan

STEP 1 – ASSEMBLE LOWER SUPPORT FRAMES:

The best trees for the construction of a tree house are apple, ash, beech, cedar, Colorado blue spruce, cypress, fir, hemlock, hickory, sugar maple, oak, and weeping willow. Attach items 1 & 2, qty. eight each to the tree trunk using item 21 lag screws qty. four per board. Use a ¼-inch drill bit and drill pilot holes to prevent wood from splitting. The elevation of the support should be 2 feet below the intended height of the tree house landing deck. As you install each board, install one screw and then check for level prior to installing the subsequent screws.

STEP 2 – ASSEMBLE MAIN SUPPORT FRAME AT EACH OF THE FOUR LANDINGS:

Assemble the main frame by first attaching item 3 to the tree trunks approximately 2 feet above the lower support. Fasten securely using item 21 lag screws, qty. eight per connection. Drill a pilot hole using a ¼-inch drill bit to prevent the wood from splitting. All subsequent beams must be installed at the same elevation. Confirm that each component is level. Continue to assemble the remaining items 4, 5, & 6 using lag screws item 21, qty. four per connection.

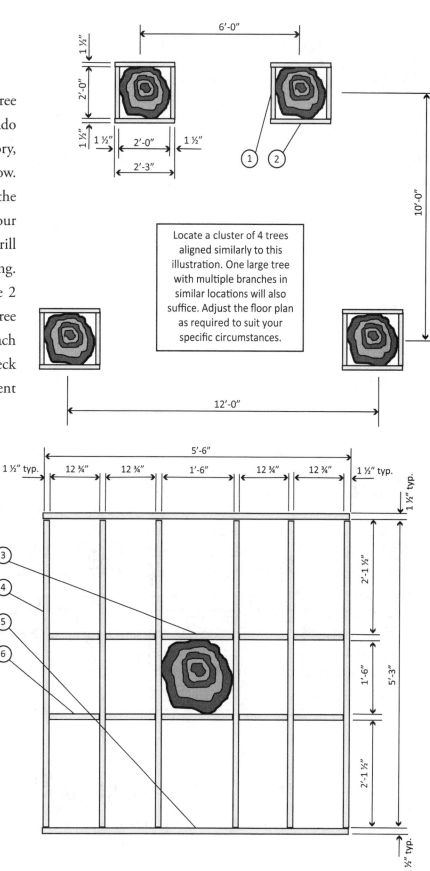

Locate a cluster of 4 trees aligned similarly to this illustration. One large tree with multiple branches in similar locations will also suffice. Adjust the floor plan as required to suit your specific circumstances.

Tree House Landings Plan

STEP 3 – INSTALL GUSSET SUPPORTS FOR EACH OF THE MAIN FRAMES:

Notch item 7 gusset at top and bottom to match lower support and upper main deck. Fasten to frame using item 21 lag bolts, qty. four per connection. Fasten steps item 10 for access to first landing only using item 21 lag bolts, qty. four per connection.

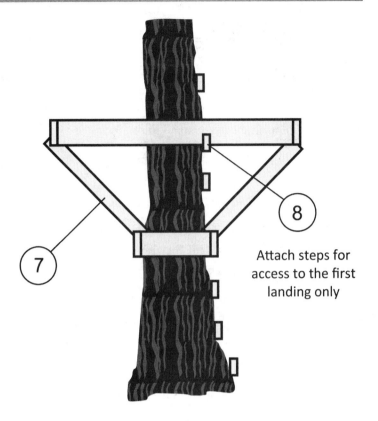

Attach steps for access to the first landing only

Landing 1

Landings 2, 3, & 4

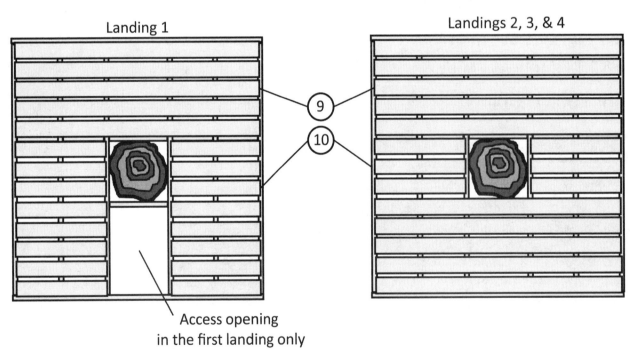

Access opening in the first landing only

STEP 4 – INSTALL DECK ON EACH OF THE FOUR LANDING FRAMES:

Attach deck boards items 9 & 10 to top of frames using 1 ⅝-inch deck screws item 22, qty. three screws per connection.

Tree House Landings Plan

STEP 5 – INSTALL HANDRAIL ON ALL FOUR LANDINGS:

Install handrail items 11, 12, 13 , & 14 using 3-inch deck screws item 23, qty. three screws per connection. Install catwalk support item 14 using lag screws item 21, qty. six screws per connection.

STEP 6 – INSTALL CATWALKS BETWEEN LANDINGS:

Attach 1-inch dia. lower rope support item 16 to main frame. Drill 1 ¼-inch hole through frame and post for a secure connection. Attach ½-inch dia. upper rope handrail item 17 to vertical post item 14. Drill ¾-inch dia. hole in upper post. Fasten steps item 18 to top side of lower rope supports. Drill ¾-inch dia. holes in steps, qty. four per board on 26-inch spacing. Secure steps to 1-inch dia. lower rope item 16 using ½-inch dia. rope item 17. Use remaining rope to tie handrail and lower rope together. Install corner benches and supports items 19 & 20 using 3-inch deck screws item 23, qty. three per connection.

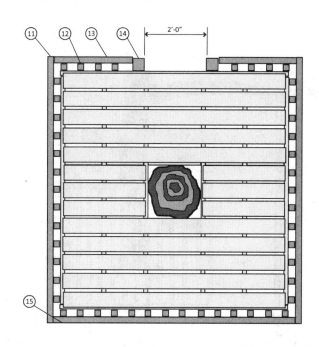

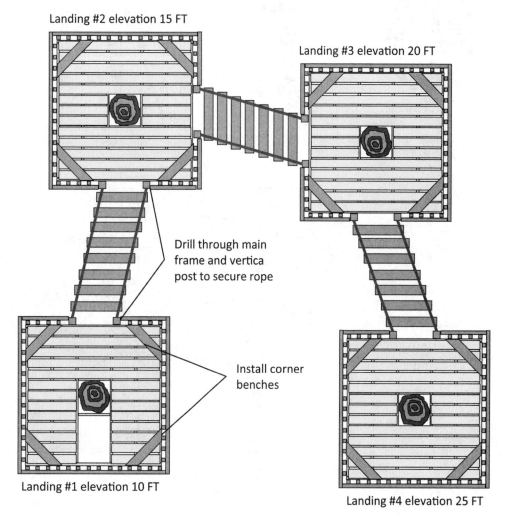

Landing #2 elevation 15 FT

Landing #3 elevation 20 FT

Drill through main frame and vertica post to secure rope

Install corner benches

Landing #1 elevation 10 FT

Landing #4 elevation 25 FT

Funhouse Tree House Plan

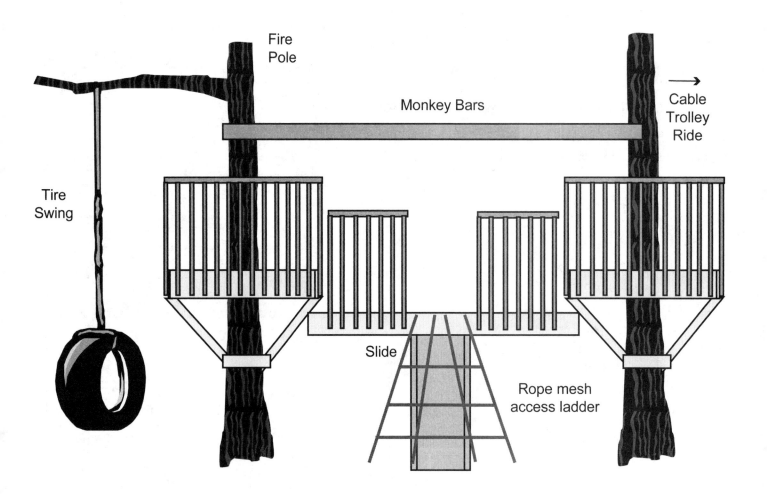

Fire
Pole

Monkey Bars

Cable
Trolley
Ride

Tire
Swing

Slide

Rope mesh
access ladder

TOOLS REQUIRED

- Safety glasses
- Gloves
- Carpenter's hammer
- 24" framing square
- 24" level
- Hand drill

- Screwdriver
- Tape measure
- Hand saw
- ¼" drill bit
- Paint brush
- ¾" wrench

Funhouse Tree House Plan

ITEM	QTY.	SIZE	DESCRIPTION	ITEM	QTY.	SIZE	DESCRIPTION
BILL OF MATERIAL							
1	7	2 x 6 x 1'-6"	Support	16	2	2 x 2 x 3'-10 ½"	Handrail
2	4	2 x 6 x 1'-9"	Support	17	1	2" dia. pipe x 10'-0"	Fire pole
3	8	2 x 6 x 3'-9"	Deck support	18	2	2 x 6 x 6'-0"	Fire pole support
4	4	2 x 6 x 4'-0"	Deck support	19	1	2 x 6 x 1'-6"	Fire pole support
5	71	2 x 2 x 3'-6"	Rail supports	20	1	10' slide	Slide
6	4	4 x 4 x 5'-6"	Mid-deck hanger	21	2	2 x 12 x 12'-0"	Monkey bar support
7	16	2 x 4 x 3'-0"	Gusset	22	9	1 ½" dia. pipe x 1'-10"	Monkey bar
8	4	2 x 10 x 8'-0"	Support	23	roll	½" dia. cable	Cable
9	3	2 x 10 x 1'-2"	Support	24	1	Cable trolley	Trolley
10	2	2 x 10 x 4'-0"	Deck	25	roll	½" dia. rope	Rope
11	12	$\frac{5}{4}$ x 6 x 14"	Deck	26	1	Large rubber tire	Tire swing
12	22	$\frac{5}{4}$ x 6 x 3'-11"	Deck	27	box	½" x 4" LG lag screws	Screws
13	2	2 x 2 x 4'-0"	Handrail	28	box	1 $\frac{5}{8}$" dia. deck screws	Screws
14	4	2 x 2 x 10 ½"	Handrail	29	box	3" deck screws	Screws
15	4	2 x 2 x 2'-6"	Handrail				

Funhouse Tree House Plan

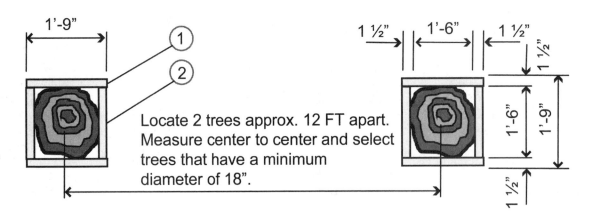

Locate 2 trees approx. 12 FT apart. Measure center to center and select trees that have a minimum diameter of 18".

STEP 1 – INSTALL LOWER SUPPORTS:

The best trees for the construction of a tree house are apple, ash, beech, cedar, Colorado blue spruce, cypress, fir, hemlock, hickory, sugar maple, oak, and weeping willow. Attach items 1 & 2, qty. four each to the tree trunk using item 27 lag screws, qty. four per board. Use a ¼-inch drill bit and drill pilot holes to prevent wood from splitting. The elevation of the support should be 2 feet below the intended height of the tree house's upper floor. As you install each board, install one screw and then check for level prior to installing the subsequent screws.

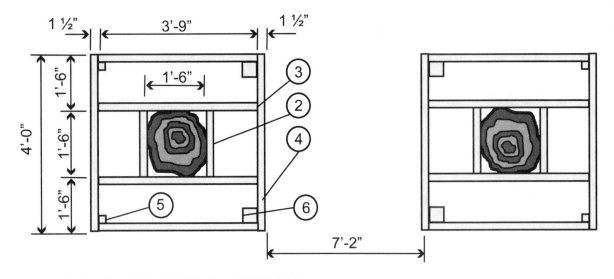

STEP 2 – CONSTRUCT UPPER PLATFORMS:

Attach supports items 2 & 3 to tree using item 27 lag screws, qty. four per connection. Assemble perimeter supports using two lag screws per connection. Install corner supports item 5 flush with the bottom of the frame and attach using 3-inch deck screws item 29, qty. four screws per connection. Install mid-deck hanger item 6 protruding 2 feet - 6 inches below bottom of frame. Attach to outer frame using lag screws item 27, qty. eight per connection. Notch gusset item 7 as shown below, and install between upper deck and lower supports typical (14) places using lag screws.

Funhouse Tree House Plan

STEP 3 – INSTALL GUSSETS:

Notch gussets item 7 to mate with upper perimeter frame and lower support. Attach item 11 using lag screws item 27, qty. two per connection.

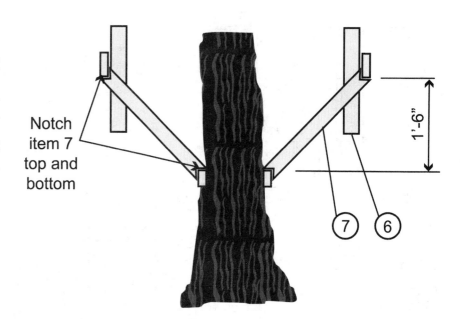

Notch item 7 top and bottom

7 6

1'-6"

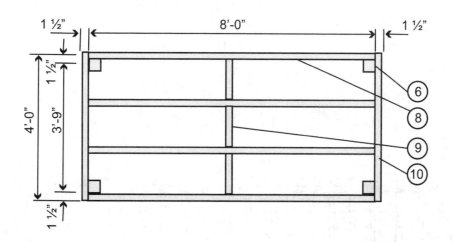

1 ½" 8'-0" 1 ½"

1 ½"

4'-0" 3'-9"

1 ½"

6
8
9
10

STEP 4 – ASSEMBLE MAIN DECK:

Attach items 4 & 8 to item 6 hanger using eight lag screws item 27 per connection. Bottom of main deck will be flush with the bottom of item 6. Assemble remaining components item 8 & 9 using two lag screws item 27 per connection.

Funhouse Tree House Plan

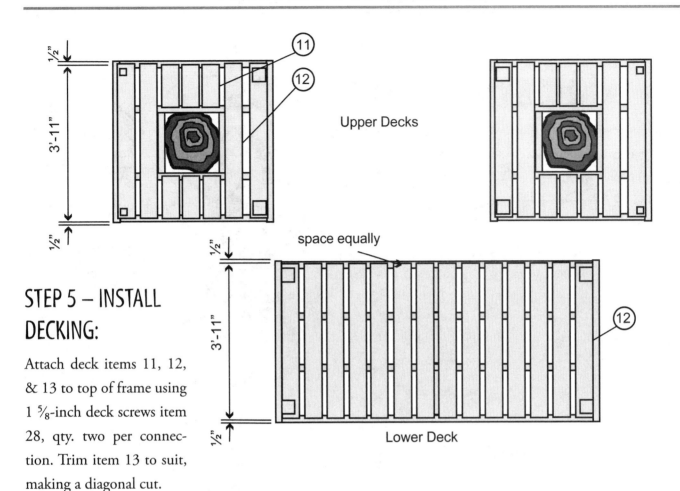

Upper Decks

space equally

Lower Deck

STEP 5 – INSTALL DECKING:

Attach deck items 11, 12, & 13 to top of frame using 1 ⅝-inch deck screws item 28, qty. two per connection. Trim item 13 to suit, making a diagonal cut.

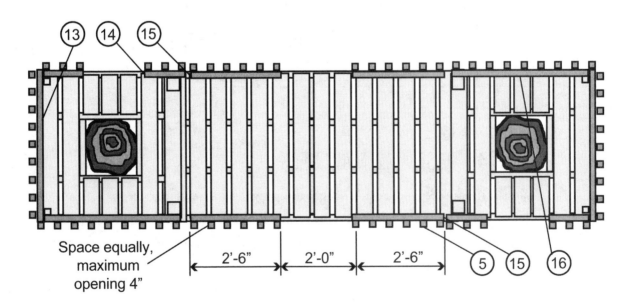

Space equally, maximum opening 4"

2'-6"   2'-0"   2'-6"

STEP 6 – INSTALL HANDRAIL AND POSTS:

Attach vertical posts item 5 to outside of frame using 3-inch deck screws item 29, qty. three screws per connection. Install top handrail items 13, 14, 15, & 16 using 3-inch deck screws item 29, qty. two per connection.

Funhouse Tree House Plan

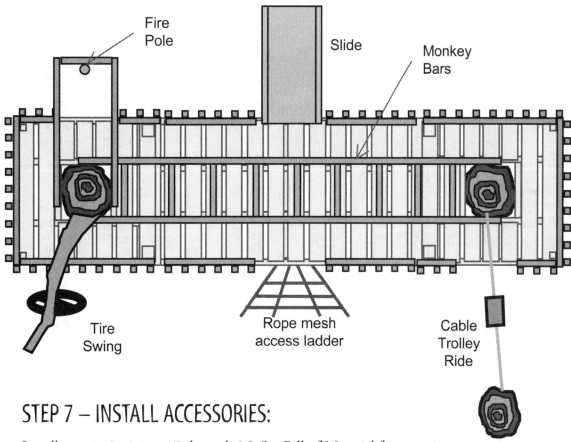

Fire Pole

Slide

Monkey Bars

Tire Swing

Rope mesh access ladder

Cable Trolley Ride

STEP 7 – INSTALL ACCESSORIES:

Install accessories items 17 through 26. See Bill of Material for quantity.

King & Queen Tree House Plan

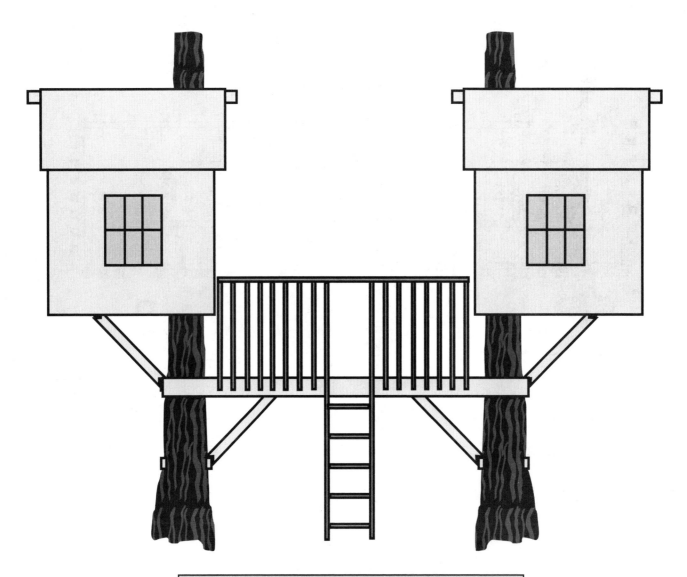

TOOLS REQUIRED	
• Safety glasses	• Screwdriver
• Gloves	• Tape measure
• Carpenter's hammer	• Hand saw
• 24" framing square	• ¼" drill bit
• 24" level	• Paint brush
• Hand drill	• ¾" wrench

King & Queen Tree House Plan

BILL OF MATERIAL							
ITEM	QTY.	SIZE	DESCRIPTION	ITEM	QTY.	SIZE	DESCRIPTION
1	4	2 x 6 x 1'-9"	Support	18	2	½" ply x 2'-0" x 6'-0"	Door
2	4	2 x 6 x 1'-6"	Support	19	4	2" hinge	Hinge
3	6	2 x 12 x 16'-0"	Deck support	20	2	Standard door latch	Latch
4	6	2 x 12 x 8'-0"	Deck support	21	4	5/4 x 6 x 3'-0"	Flower box
5	12	2 x 12 x 1'-5 ¼"	Deck support	22	4	5/4 x 6 x 6"	Flower box
6	23	5/4 x 6 x 7'-11"	Deck boards	23	8	2'-0" x 3'-0" window	Window
7	12	2 x 12 x 7'-9"	Deck support	24	2	2 x 4 x 10'-0"	Ladder
8	8	2 x 6 x 3'-6"	Gusset	25	12	2 x 4 x 1'-6"	Ladder rungs
9	26	½" ply x 4'-0" x 8'-0"	Deck	26	4	2 x 4 x 3'-0"	Ladder
10	8	2 x 4 x 8'-0"	Wall stud	27	34	2 x 2 x 4'-0"	Handrail
11	24	2 x 4 x 7'-9"	Wall stud	28	1	2 x 4 x 12'-0"	Handrail
12	24	2 x 4 x 14 ¼"	Wall stud	29	2	2 x 4 x 5'-0"	Handrail
13	2	2 x 8 x 10'-0"	Roof truss	30	box	½" x 4" lag screws	Screws
14	28	2 x 6 x 6'-0"	Roof truss	31	box	1 5/8" deck screws	Screws
15	20	2 x 4 x 10'-0"	Wall stud	32	box	3' deck screws	Screws
16	4	2 x 4 x 7'-5"	Wall stud	33	bdl.	Standard roofing shingles	Shingles
17	16	2 x 4 x 1'-8 ¼"	Wall stud	34	box	1" roofing nails	Nails

King & Queen Tree House Plan

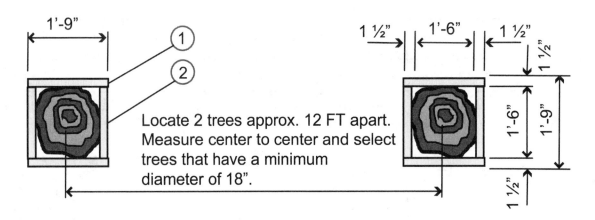

Locate 2 trees approx. 12 FT apart. Measure center to center and select trees that have a minimum diameter of 18".

STEP 1 – INSTALL LOWER SUPPORTS:

The best trees for the construction of a tree house are apple, ash, beech, cedar, Colorado blue spruce, cypress, fir, hemlock, hickory, sugar maple, oak, and weeping willow. Attach items 1 & 2, qty. four each to the tree trunk using item 30 lag screws, qty. four per board. Use a ¼-inch drill bit and drill pilot holes to prevent wood from splitting. The elevation of the support should be 2 feet below the intended height of the tree house deck. As you install each board, install one screw and then check for level prior to installing the subsequent screws.

STEP 2 – INSTALL DECK FRAME:

Attach items 2, 3, & 4 to tree using item 30 lag screws, qty. six per connection. Assemble remaining frame items 2, 3, 4, & 5 using lag screws item 30, qty. three per connection.

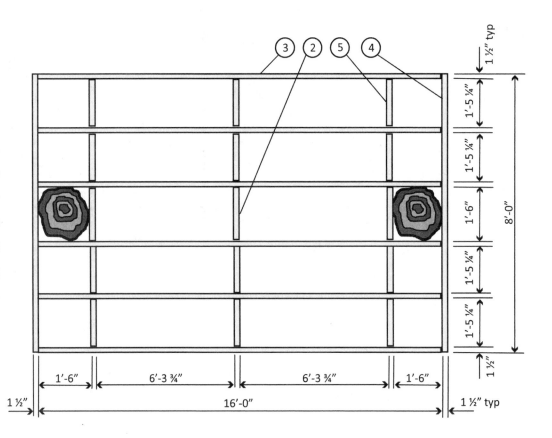

King & Queen Tree House Plan

STEP 3 – INSTALL DECK BOARDS:

Attach item 6 deck board to frame using item 31 deck screws 1 ⅝-inch long, qty. three per connection.

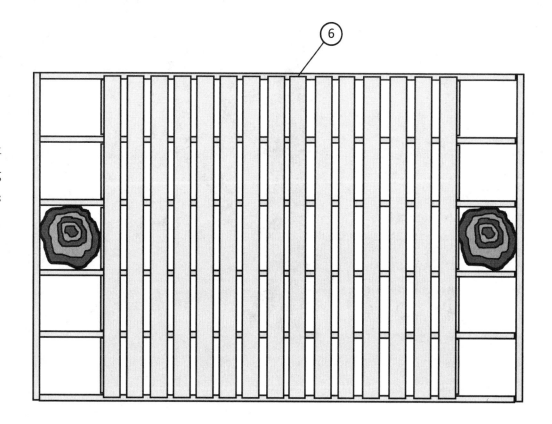

STEP 4 – INSTALL UPPER FLOOR SUPPORTS, X (2):

Attach items 2, 3, & 7 to tree using item 30 lag screws, qty. six per connection. Assemble remaining frame items 2, 3, 5, & 7 using lag screws item 30, qty. three per connection.

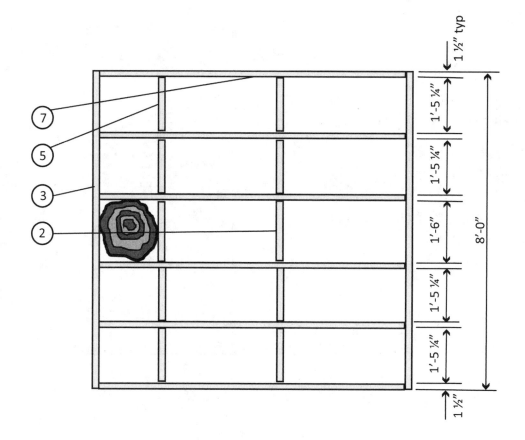

King & Queen Tree House Plan

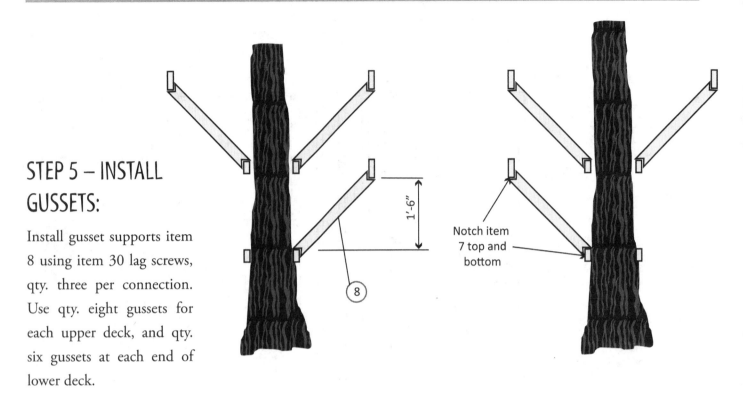

STEP 5 – INSTALL GUSSETS:

Install gusset supports item 8 using item 30 lag screws, qty. three per connection. Use qty. eight gussets for each upper deck, and qty. six gussets at each end of lower deck.

STEP 6 – INSTALL UPPER FLOOR DECK, X (2):

Attach upper deck to upper frame using 1 ⅝-inch deck screws item 31 on 6-inch spacing.

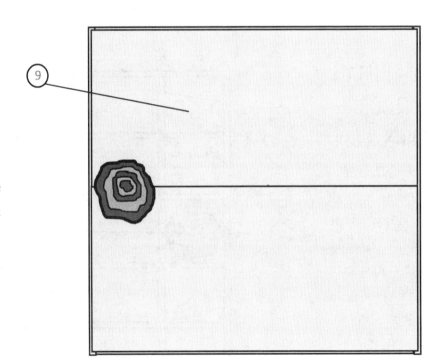

King & Queen Tree House Plan

STEP 7 – ASSEMBLE SIDE WALLS, QTY. FOUR:

Assemble side wall frame items 11 & 12 using 3-inch deck screws item 32, qty. two per connection. Attach side wall item 9 to frame using 1 ⅝-inch deck screws item 31 on 6-inch spacing. Install assembly on top of upper deck using item 30 lag screws on 6-inch centers.

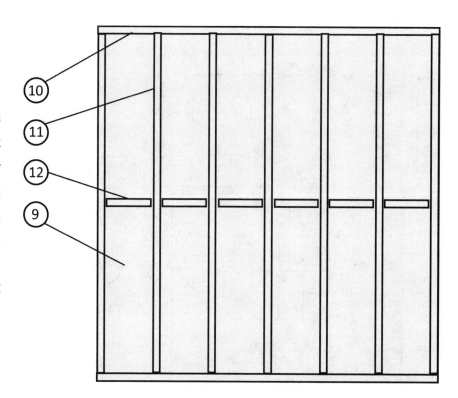

STEP 8 – ASSEMBLE FRONT AND REAR WALLS, X (4):

Assemble end walls frame items 15, 16, & 17 using 3-inch deck screws item 32, qty. two per connection. Attach wall item 9 to frame using 1 ⅝-inch deck screws item 31 on 6-inch spacing. Install assembly on top of upper deck using item 30 lag screws on 6-inch spacing. Install roof truss assembly items 13 & 14 on 16-inch centers using lag screws item 30.

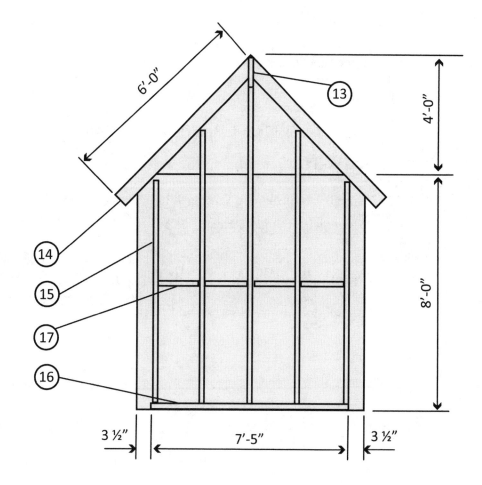

King & Queen Tree House Plan

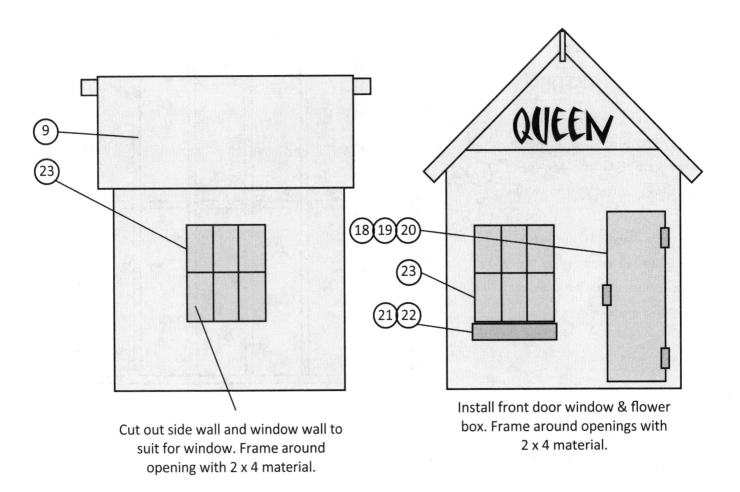

Cut out side wall and window wall to
suit for window. Frame around
opening with 2 x 4 material.

Install front door window & flower
box. Frame around openings with
2 x 4 material.

STEP 9 – INSTALL WINDOWS IN SIDE WALLS & BACK WALL, QTY. SIX:

Install roof deck item 9 using 1 ⅝-inch deck screws item 31 on 6-inch centers
Cover with shingles items 33 & 34.

STEP 10 – INSTALL DOOR & FRONT WINDOW:

Install door, front window, and flower box.

King & Queen Tree House Plan

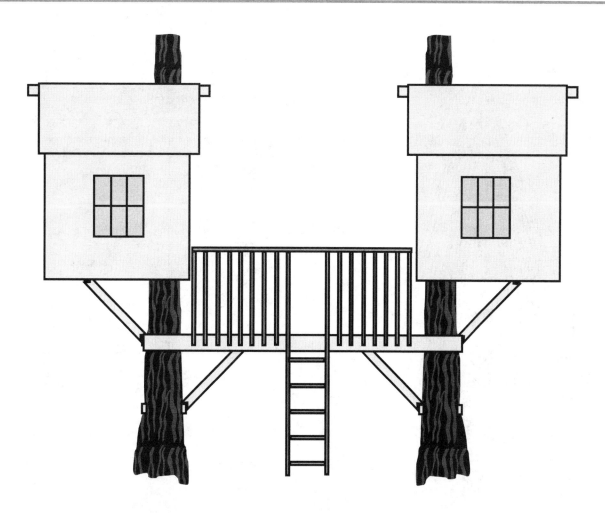

STEP 11 – INSTALL LADDERS & HANDRAIL:

Install lower ladder items 24 & 25 using 3-inch deck screws item 32, qty. four per connection. Assemble upper ladders to provide access to doors using item 26 & 26 and deck screws item 32, qty. four per connection. Attach handrail items 27, 28, & 29 to deck using 3-inch deck screws item 32, qty. three per connection.

Teepee Tree House Plan

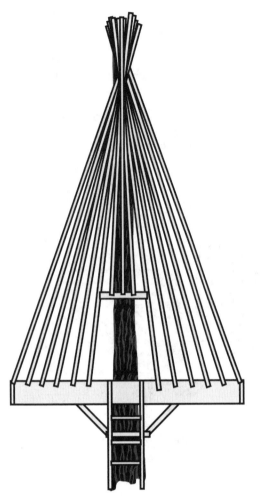

TOOLS REQUIRED

- Safety glasses
- Gloves
- Carpenter's hammer
- 24" Framing square
- 24" level
- Hand drill

- Screwdriver
- Tape measure
- Hand saw
- $\frac{3}{16}$" drill bit
- Paint brush
- $\frac{9}{16}$" wrench

BILL OF MATERIAL

ITEM	QTY.	SIZE	DESCRIPTION	ITEM	QTY.	SIZE	DESCRIPTION
1	2	2 x 6 x 1'-9"	Support	11	44	2 x 2 x 16'-0"	Wall
2	6	2 x 6 x 1'-6"	Support	12	4	2 x 2 x 12'-0"	Wall
3	3	2 x 12 x 4'-0"	Deck support	13	1	2 x 4 x 3'-0"	Wall
4	8	2 x 12 x 1'-5 ¼"	Deck support	14	2	2 x 4 x 4'-0"	Ladder
5	4	2 x 12 x 4'-3"	Deck support	15	3	2 x 4 x 1'-6"	Ladder
6	2	2 x 12 x 7'-0"	Deck boards	16	8	2 x 6 x 3'-6"	Gusset
7	4	2 x 12 x 1'-0"	Deck support	17	box	½" x 4" lag screws	Screws
8	2	2 x 12 x 12'-0"	Deck support	18	box	1 ⅝" deck screws	Screws
9	2	2 x 12 x 1'-1 ½"	Deck support	19	box	3' deck screws	Screws
10	4	½" ply x 4'-0" x 8'-0"	Deck				

Teepee Tree House Plan

STEP 1 – INSTALL LOWER SUPPORT:

Locate a tree with a trunk diameter of approximately 18 inches minimum. The best trees for construction of a tree house are apple, ash, beech, cedar, Colorado blue spruce, cypress, fir, hemlock, hickory, sugar maple, oak, and weeping willow. Attach items 1 & 2, qty. two each, to the tree trunk, using item 17 lag screws, qty. four per board. The elevation of the support should be 2 feet below the intended height of the tree house floor. As you install each board, install one screw and then check for level prior to installing the subsequent screws.

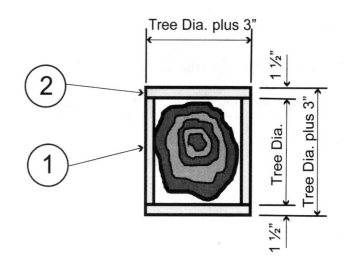

STEP 2 – INSTALL LOWER-LANDING PERIMETER FRAME:

Attach inner supports items 2 & 8 to the tree trunk, using item 17 lag screws, qty. six per board. The elevation of the support should be 2 feet above the lower support. As you install each board, install one screw, and then check for level prior to installing the subsequent screws. Install the outer supports items 3, 4, 5, 6, 7, 8, & 9 using item 17 lag screws, qty. three per connection.

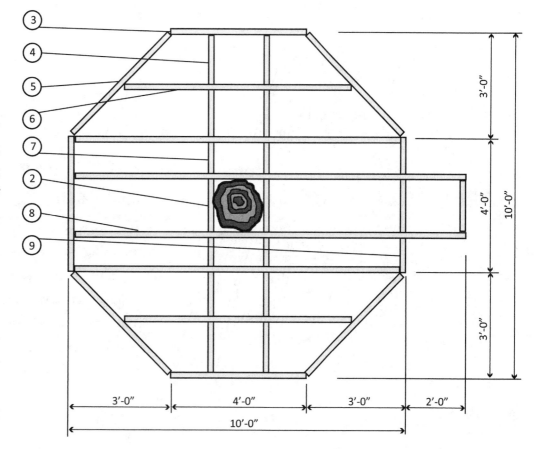

Teepee Tree House Plan

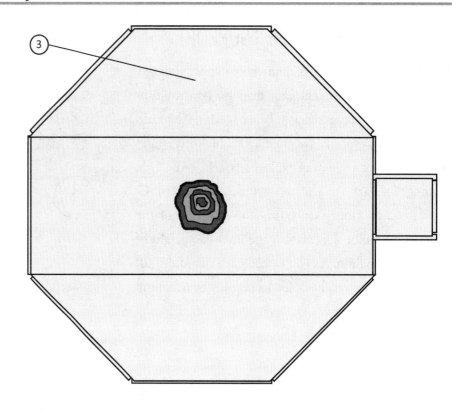

STEP 3 – INSTALL DECK:

Cut plywood item 10 to suit. Attach to deck using 1 ⅝-inch deck screws item 18 on 6-inch spacing.

STEP 4 – ASSEMBLE TEEPEE WALLS & LADDER:

Attach teepee walls items 3, 4, & 5 to deck and tree using 3-inch deck screws item 19, qty. two per connection. Assemble ladder items 8 & 9 using 3-inch deck screws item 19, qty. four per connection. Notch gussets item 16 to suit and attach using lag screws item 17, qty. two per connection.

Dueling Slides & Swinging Bridge Tree House Plan

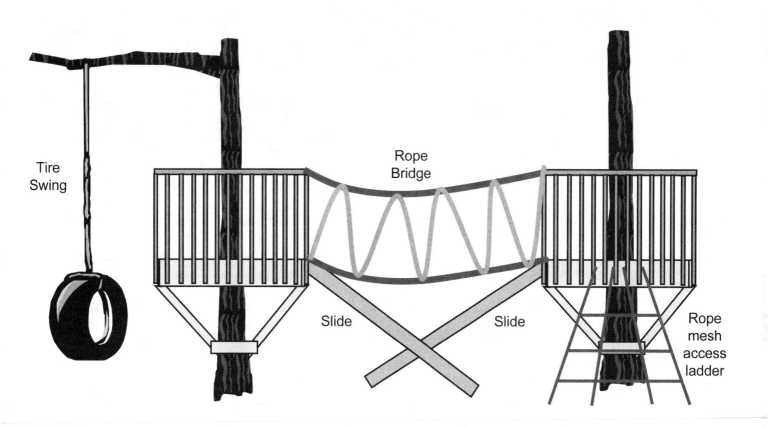

Tire
Swing

Rope
Bridge

Slide

Slide

Rope
mesh
access
ladder

TOOLS REQUIRED	
• Safety glasses	• Screwdriver
• Gloves	• Tape measure
• Carpenter's hammer	• Hand saw
• 24" Framing square	• ³⁄₁₆" drill bit
• 24" level	• Paint brush
• Hand drill	• ⁹⁄₁₆" wrench

Dueling Slides & Swinging Bridge Tree House Plan

ITEM	QTY.	SIZE	DESCRIPTION	ITEM	QTY.	SIZE	DESCRIPTION
1	7	2 x 6 x 1'-6"	Support	13	12	2 x 6 x 2'-0"	Bridge steps
2	4	2 x 6 x 1'-9"	Support	14	roll	1" rope	Bridge
3	8	2 x 6 x 3'-9"	Deck support	15	1	2" dia. pipe x 10'-0"	Fire pole
4	4	2 x 6 x 4'-0"	Deck support	16	2	2 x 6 x 6'-0"	Fire pole support
5	8	2 x 2 x 3'-6"	Rail supports	17	1	2 x 6 x 1'-6"	Fire pole support
6	16	2 x 6 x 3'-6"	Gusset	18	2	10' slide	Slide
7	12	¾ x 1'-1 ½"	Deck	19	roll	½" dia. rope	Rope
8	8	¾ x 3'-11"	Deck	20	1	Large rubber tire	Tire swing
9	48	2 x 2 x 3'-6"	Handrail	21	box	½" x 4" LG lag screws	Screws
10	4	2 x 2 x 1'-0"	Handrail	22	box	1 ⅝" dia. deck screws	Screws
11	4	2 x 2 x 4'-0"	Handrail	23	box	3" deck screws	Screws
12	4	4 x 4 x 4'-0"	Bridge supports				

Table title: BILL OF MATERIAL

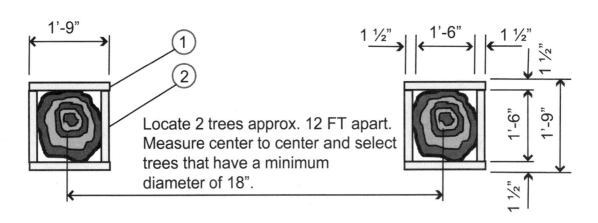

Locate 2 trees approx. 12 FT apart. Measure center to center and select trees that have a minimum diameter of 18".

STEP 1 – INSTALL LOWER SUPPORTS:

The best trees for the construction of a tree house are apple, ash, beech, cedar, Colorado blue spruce, cypress, fir, hemlock, hickory, sugar maple, oak, and weeping willow. Attach items 1 & 2, qty. four each to the tree trunk using item 21 lag screws, qty. four per board. Use a ¼-inch drill bit and drill pilot holes to prevent wood from splitting. The elevation of the support should be 2 feet below the intended height of the tree house's upper floor. As you install each board, install one screw and then check for level prior to installing the subsequent screws.

Dueling Slides & Swinging Bridge Tree House Plan

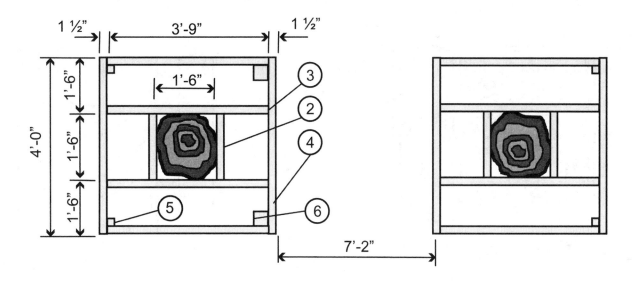

STEP 2 – CONSTRUCT UPPER PLATFORMS:

Attach supports items 2 & 3 to tree using item 21 lag screws, qty. four per connection. Assemble perimeter supports using 2 lag screws per connection. Install corner supports item 5 flush with the bottom of the frame and attach using 3-inch deck screws item 23, qty. four screws per connection. Notch gusset item 6 as shown below and install between upper deck and lower supports typical (14) places using lag screws.

STEP 3 – INSTALL GUSSETS:

Notch gussets item 6 to mate with upper perimeter frame and lower support. Attach item 11 using lag screws item 27, qty. two per connection.

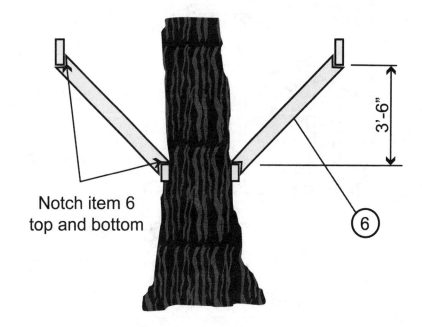

Notch item 6 top and bottom

Dueling Slides & Swinging Bridge Tree House Plan

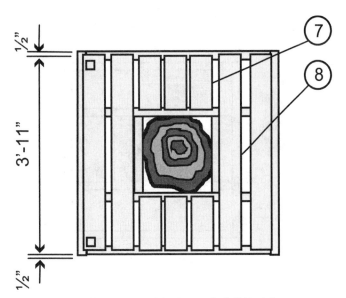

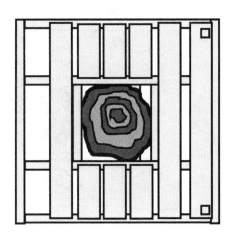

STEP 4 – INSTALL DECKING:

Attach deck items 7 & 8 to top of frame using 1 ⅝-inch deck screws item 22, qty. three screws per connection.

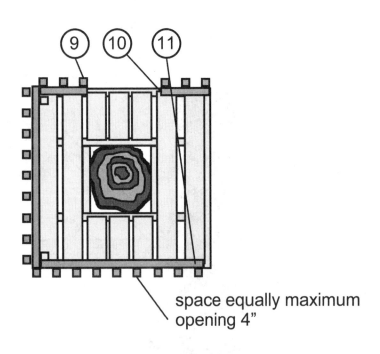

space equally maximum opening 4"

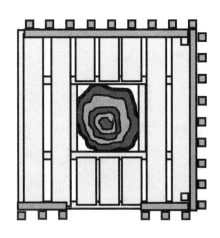

STEP 5 – INSTALL HANDRAIL AND POSTS:

Attach handrail items 9, 10, & 11 to top of deck using 3-inch deck screws item 23, qty. two per connection.

Dueling Slides & Swinging Bridge Tree House Plan

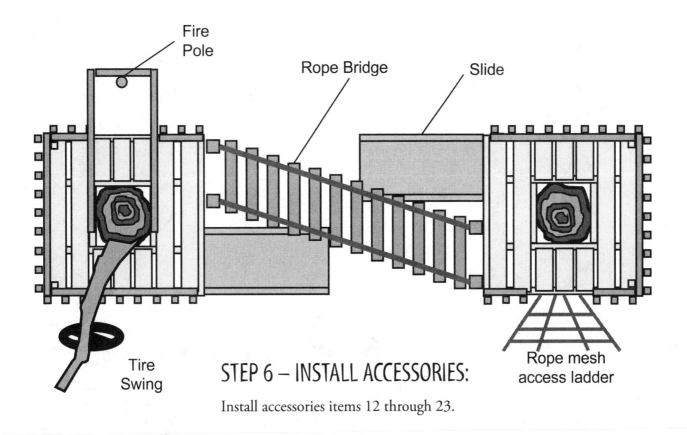

Fire
Pole

Rope Bridge

Slide

Tire
Swing

Rope mesh
access ladder

STEP 6 – INSTALL ACCESSORIES:

Install accessories items 12 through 23.

This tree house was designed by Barbara Butler.

Conclusion

What is it about tree houses that fascinates and delights people? Could it be something about being suspended high above the world below? Do tree houses stimulate play and creativity as no other structure can? Could it be the illusion of escaping Earth, into another world? No matter the reason that people decide to build tree houses, they afford privacy and the opportunity for a deeper encounter with nature — and with one's self.

Tree houses are growing in popularity and in numbers. As outlined in this book, it is possible to build a functional tree house with the barest of materials — including materials you can find and reclaim. And it is possible to build elaborate and expensive tree houses with all the comforts of a ground house. Tree houses can be designed to accommodate adults' fantasy of tree-top luxury living, or be designed to give a child a place where he or she can use their imagination. There are no limits to what you can do with the right materials and a great plan.

This book presents a range of tree houses, from the simple to the extravagant, so that tree house builders can borrow ideas and draw inspiration from others' work. In this book, we have covered everything from basic tree house building to building with a specific, more elaborate plan. The information presented will assist you as you progress through the building process.

Remember to:

- Use the tips provided to help you pick the right trees for your tree house.

- Bring the list of tools provided in this book with you when you go to the hardware store. A printable version is available on the companion CD-ROM.

- Look through all the pictures and read the case studies for inspiration.

- Check with your town, county, and state's regulations before building a tree house.

- Make sure your tree house is strategically placed so that it does not invade the privacy of your neighbors.

- Read over the safety section before you start building.

Good luck with this project. Remember, this can be something for the whole family to participate in. Ask everyone for their opinion, especially during the decorating stage. This experience will be one that you and your family never forget. Someday, when your children are older, they will remember the time the family built a tree house, and they may build one for their own children in the future. Do not forget to take pictures along the way; every moment is a moment that should be captured and cherished. Have fun and be safe.

Appendix:
More Case Studies
Experiences From the Professionals

Case Study: Ray Cirino's Innovative and Modern Structures

Ray Cirino

Owner/designer of Permaquest

Venice, California

Ray Cirino, whose ultra-modern work has been featured in the Los Angeles Times and in LA Weekly, designs and builds tree houses in teardrop shapes with the goal of making them appear like tulips emerging from the earth. "I believe in mimicking nature, since it has been around for a billion years and knows what is strong and beautiful," he said.

Cirino, who has been building tree houses since 1967, estimates that his total number of completed projects is 11. His tree houses, which have sold in a range of prices from $5,000 to $110,000, are built from salvaged material, as well as found and recycled objects (such as driftwood and discarded rocket parts). In one of Cirino's futuristic tree houses, which has upturned corner support beams, the railings are fashioned from curved pipes he found in a movie studio scrap-heap in Hollywood. He views tree houses as teaching tools to spread the word about bio-sustainability and living within the natural parameters.

One of his many innovations is a unique means of suspension that does not involve putting holes in the tree. Cirino encircles the tree with a thick rubber collar, then wraps that in an adjustable, stainless-steel mesh into which rebar is inserted to support the platform. Another innovative design is a tear-shaped, personal outside dwelling (POD), which he describes as "a functional yurt with insulation." PODS are light but sturdy — constructed of shaped 1/4-inch plywood that makes the structure — and ideal for warmer climates like Southern California. The units are made with 2-inch foam insulation, but simply by adding a higher-quality insulation, they could be serviceable in colder climates. In addition to their current use as a backyard studios and detached bedrooms, they also work as tree houses.

Cirino feels that a tree house offers children a "nest" where they feel safe. "The perfect tree house will ultimately educate children and make them feel they have a safe nest," he said. "Tree houses give children the perspective of being high and looking down on the Earth and their lives. Children feel empowered."

 See the color insert for photos of Ray Cirino's creations.

Case Study: An Architect-Designed Tree House in Michigan

Michael Poris

AIA, McIntosh Poris Associates

Birmingham, Michigan

A three-tree solution to a client's request for a private retreat in the woods of suburban Detroit presented a real challenge to architect Michael Poris — plus, the added fact that the client was also a friend. The friend/client wanted a retreat built on his wooded 5-acre site in West Bloomfield, Michigan, as a place to de-stress from his job as an advertising executive. He wanted the tree house to be free of electronic hookups; a place where he could meditate, read, write, and spend time with his two young sons.

Poris designed and then helped build a 600-square foot tree house supported by three pine trees, 26 feet off the ground on a sloping hillside with a panoramic view of the surrounding countryside. At first, he admitted, "it did not seem a very tree house-friendly site," but selecting the right combination of trees on the right hill made all the difference.

The pine trees support the tree house by a system of wood beams reinforced with metal cable, and a fourth tree had a series of stairs and platforms that led up to the entrance. The 10- by 12-foot structure rested on a 12- by 26-foot platform that left an additional 14- by 12-foot terrace deck. One-third of the total area was enclosed, and the other two-thirds were open, according to the architect.

Services of an arborist enlisted

"Getting the platform to connect to three trees, 26 feet in the air, required the services of an arborist," Poris said. "There was a person in our architectural office that did these kinds of things. He climbed into the trees and put in these three thick bolts to support the platform. He also installed them so that as the trees grow, you can loosen the bolts."

"Once those supporting bolts were in the trees, we had a carpenter build three trusses made from tree limbs that were 20 to 30 feet long. We then connected the platform with wires to the three trees and used big chains attached to the bolts and to the beams to suspend the platform. We had to come up with something that would move in the wind," he said.

City officials told Poris that he did not need a permit to build the house, and he had a structural engineer make an analysis to see if it would work. The crew used a Jeep to lift the trusses up in place. Only 40 feet from the main house, the tree house was warmed by electric heat, and guests could use the bathrooms in the main house.

A downsloping hill in front of the tree house gave the sensation of being very high off the ground.

"You feel like you are in the mountains," Poris said. The inside is all cedar with a metal roof, skylights, and windows on all sides.

Somewhat like a Japanese teahouse, all four sides of the tree house can be opened. Three sides have large casement windows, and the fourth side has windows that also serve as doors. All the windows can be opened, with or without screens.

 See the color insert for more photos of Michael Poris' tree houses.

Case Study: A Tree House Built with Recycled Lumber and Love

John Farless

Communications director

Evansville, Indiana

Inspired — perhaps "driven" might be a better word — by his 5-year-old daughter Brenn, John Farless of Evansville, Indiana, set about building her a tree house with free, recycled lumber and several how-to books under his arms.

Although Farless, a publicist with Saint Meinrad Archabbey and School of Theology, had never before built a tree house, he dug out lots of material by using the Internet and relied on his brother, who had roofing experience, to conceive and execute the project.

"The most challenging part was just getting my head wrapped around the tree house concept," he said.

Although he had roughed out a sketch of what he wanted the tree house to look like, Farless said the ultimate shape depended to a large degree on the recycled materials he was able to use. He posted an item on **www.freecycle.com** under "scrap lumber" and hit the jackpot.

"One guy who is a builder/contractor had tons of scrap wood and lumber piled up," he said. "He gave us enough to build a tree house for my daughter, and one for my brother's son, too. I was basing my design on what materials we had on hand, but this made it possible to greatly improve the design."

And Brenn had her own ideas, too, as well as her own tool kit. "She hangs out while we were working," her father said. "She had definite ideas about where she wanted everything to be. She can drive a nail like you would not believe."

Basically, the tree house rests on an 8-square-foot platform elevated 6 feet off the ground. The structure rests on two 6- by 6-foot posts in front and is built around a large maple tree. Most of the recycled lumber he used was pressure-treated to withstand weather and seasonal changes.

To hold everything together and to the tree, Farless used 8-inch lag bolts with 2- by 8-inch beams and 2- by 6-inch braces, including one knee brace. He researched the Internet to be sure that lag bolts would not injure the tree, he said.

Farless said he grew up on a farm where he was responsible for many kinds of chores, including some carpentry. That experience, plus the help of his brother and his research, have brought the tree house to life.

Among the pieces of scrap wood he found in the builder's recycled pile was a wooden arch that had served as a template for setting a round concrete pad. He decided it would make a great arch rising from the back of the tree house and sloping forward. He planned to cover it with roofing shingles — provided he received Brenn's approval.

Case Study:
Natural Beauty in Big Sur Tree Houses

Mickey Muennig

Post Ranch Innn • **www.postranchinn.com**

Big Sur, California

As an architect, Mickey Muennig had designed practically everything under the sun — except a tree house — when he was commissioned to build seven tree houses as part of the Post Ranch Inn in Big Sur, California.

But that proved to be a small obstacle, once he had offered his "Robinson Crusoe takeoff" concept to the owners of Post Ranch Inn, a 40-unit luxury retreat on 100 acres of dramatic Big Sur coastline high above the Pacific Ocean. The triangular tree house design he came up with was replicated six times for a total of seven tree houses.

Because of construction restraints by the California Coastal Commission, the tree houses could not exceed 450 square feet, but in that space, the renowned architect was able to fit in a wet bar and small refrigerator with large picture windows that offer sweeping views of the Pacific and surroundings.

"The design combines 10-inch beams that are about 9 feet off the ground, set in a hillside with thick stands of live oak and redwood," he said. "The support poles go 10 feet into the ground and rest on concrete, on a sloping hillside."

Each tree house was placed so construction would not cut the trees or root systems, and focused on maximizing views, according to Muennig. Support for the tree houses is fairly conventional, with no floating platforms.

The tree house design was completed in 1988, and the seven structures were opened in 1992. "We used No. 1 clear heart structural redwood for the beams," Muennig said. They worked with a variety of wall and flooring surfaces, including ceramic tile bathtubs, rich redwood paneling, and a combination of wall-to-wall carpeting and slate floors.

In keeping with the environmentally sensitive design of the tree houses and other structures at Post Ranch Inn, the resort announced in May 2009 that it had switched on a 990-panel solar installation to provide electrical energy. It is the largest hotel solar project of its kind in California, according to managing partner Mike Freed.

The 210-watt photovoltaic solar panels produce about 400,000 kWh of pollution-free electricity. The panel is designed to maximize its efficiency by following the sun's movement throughout the day.

After the tree houses were built in Big Sur, Muennig was asked by the San Francisco Arboretum to build a tree house. He came up with a simpler design like the Post Ranch Inn tree house, built it in sections, and carried it to the arboretum, where it was assembled.

Muennig, who moved to Big Sur in 1977 from Denver, recently retired, but his partner, Tim Brattan, continues the business.

 See the color insert for more photos of Post Ranch Inn.

Case Study:
A Father and Son Inter-Generational
Tree House Project

Scott Daves

CEO, Scott Daves Construction

Cary, North Carolina

As a professional builder of expensive homes in Cary, North Carolina, Scott Daves took advantage of the recession to share the magic and fun of building a tree house with his 10-year-old son, Chad.

When Daves was 11, he built a three-story tree house near his home in Winston-Salem, complete with carpet and electricity. The local newspaper published a front-page feature story about his accomplishment, which he said started his career as a builder.

Before building his headline-grabbing tree house, Daves had built many tree houses, each gradually becoming more elaborate since the first time he nailed "two or three boards across some limbs." But he clearly remembers the hours he spent on that perch, daydreaming in a kind of reverie.

During his college years, Daves worked on building sites whenever he had the opportunity, and he started his own construction business in 1999. He usually builds eight to ten homes a year, but a while ago, his son noticed the newspaper article on the wall of his office, and asked when they could build their own tree house.

Daves and his son used a computer-assisted design program to design their tree house. "[Chad] saw how you first get an idea, draw a plan, estimate costs, and then build," Daves said. "He got an apprenticeship in construction."

Father and son chose a white oak tree of about 20 inches in diameter for their construction site. They built a rectangular platform secured by lag bolts and assembled with screws using a screw gun. The tree house is about 20 feet off the ground on the high side and 15 feet on the lower side.

"We put a pulley in the tree with a long rope, and I had Chad hoist the pieces up and hold them in place," said Daves. "Then I got the screws started, and he finished them with the screw gun." They eventually painted the structure brown, so it would blend in with the natural surroundings. Total construction time was probably about four or five days, stretched over a two-month time period.

Daves and son also installed a trap door and added a trampoline nearby. The whole enterprise is about freeing and nurturing the imagination, Daves said.

"When Chad comes home, he goes straight to the tree house," he said. "If he's like me, I think it makes him feel higher and contemplative. He's thrilled about it. Having him just sit there is something special."

Daves recalls a poster his father gave him years ago — a sweeping landscape of the Smoky Mountains in fall, and a little tree house in the near distance with a rope hanging down. Its inscription reads: "Remember the Good Times."

Father and Son: *Scott Daves (left) and Chad Daves enjoy a quiet moment looking over the countryside from their tree house in North Carolina.*

Case Study:
When a Tree House Strangles its Supporting Tree

Stephen Roe, patent attorney

Lathrop and Clark LLP

Madison, Wisconsin

Sometimes, rescuing a tree house is preferable to building one, especially when the tree house is terrific except for one flaw — strangling the trees.

In what amounts to a giant "oops," the builders of a tree house that came with the property Steve Roe purchased in 1997 in Alexandria, Virginia, had failed to provide sufficient space for an old maple tree with three trunks to grow and expand.

"It was incredibly well-constructed, but the builder made no allowance for the growth of the supporting maple tree," said Roe, who is also a mechanical engineer. "The original tree house platform was hexagonal and was built around a maple tree whose main trunk divided at about 3 feet into daughter trunks. The original roof was also hexagonal but was built by tightly encircling the larger and more vertical daughter trunk with lengths of 2- by 4-inches with 2- by 4-inch joists that extended from the 2x4 sections, attached to the trunk by posts that extended up from the corners of the platform."

Roe said plywood was laid over joists to create the platform, which was covered with regular house shingles. In addition, the tree house was supported by 2- by 6- by 8-inch beams bolted into each of the three trunks. The octagonal structure above the platform was built to resemble a Tiki hut, Roe said. The roof was constructed of 2x4s at a height of 5 feet, with a skylight.

Excellent, with one fatal flaw

"It was great construction — without any consideration at all to the tree," he said. Within a couple of years of moving into the house, the maple's leaves were few and droopy-looking. Roe's wife observed, "The tree house is killing the tree." Roe noticed that the leaves on one maple trunk were yellow, while the other maples were sprouting healthy, green leaves.

"As the tree grew, there was no room left for the sections and joists to be displaced away from the tree as the diameter of the trunk expanded, thus strangling that trunk," Roe said.

He set about trying to rescue the tree. He made an Internet search for any information he could find about tree houses, their construction, and their maintenance.

Although Roe had not built a complete tree house from scratch, when he was about 11 years old, he and a friend came across remnants of some kind of tree structure in suburban Chicago. The two boys scrounged pieces of wood from their fathers and "sort of rebuilt" the decaying tree house.

He grew up in a rural, 2-acre-lot sized area. Whenever his father was home, he remained focused on taking care of adult chores rather than building a tree house, according to Roe.

Rebuilding house to rescue it

First, he tried expanding the tree holes by 4 inches and provided more under-bracing to support the platform. Then he removed the end pieces of the plywood to further relieve stress on the tree.

"I ended up substantially rebuilding the tree house, including putting a new floating roof on so it did not strangle the tree, and adding a new wing of 4- by 8-foot plywood decking to provide a sleeping loft. I replaced the floor and its supports to relieve stress and deformation caused by tree growth," he said.

He removed the two beams that supported the roof and cut four replacement beams from 2- by 4-inches by 10-feet and mounted them between a gap in the trees to further relieve stress on the ailing maple trunk. The new roof floats on a roof ridge, relative to the trees, suspended by the new beams.

In doing so, he also left a 2-inch gap between the roof and the tree to allow room for growth and expansion. When he rebuilt the roof, he ran the shingles to the tree trunk, knowing they could always be removed later.

By the time his family moved from Virginia to Wisconsin, the maple tree had fully recovered.

"Recovery started in the spring, but by summer the tree already looked better," he said. "Everybody said it was the coolest thing they had ever seen."

Case Study:
A Special Tree House for Special Kids of All Ages

Camp Matz Tree House,

Watertown, Wisconsin

What kid would not love to climb up into a tree house, act out imaginary dramas high in the treetops, then zoom back to earth on a zip line or down a fireman's pole? Until recently, this dream had been inaccessible to handicapped and developmentally disabled children.

"This is neat!" exclaimed one camper upon entering the Camp Matz Tree House in Watertown, Wisconsin. "I have never been in a tree house before."

The tree house was opened in 2004 as an adjunct to the roster of services and activities offered through Bethesda Lutheran Communities at Camp Matz, on its 440-acre site in Watertown. Bethesda operates "six or eight" group homes in Wisconsin, according to Reuben Schmitz, corporate director of Development Operations. Funding of approximately $135,000 for construction came from a private donation, while nonprofit Forever Young Treehouses in Vermont takes care of the necessary legal paperwork for building permits and maintains an on-site engineer during construction, Schmitz said. Laborers for Christ, an organization of retired builders who work for minimum wage, provided much of the labor at minimum wage.

Camp Director Donna Winter proudly said that Camp Matz has the first wheelchair-accessible tree house in the Midwest. Added Schmitz: "The whole point is for the campers to mesh with nature."

Although campers range in capabilities from profoundly to mildly disabled, those who are aware of their surroundings are usually infatuated with the fact they are in a structure 15 to 20 feet off the ground, built around trees. "Some of these folks have not had much opportunity to connect with nature," Winter said.

Each week during summer camp, a new group of 15 to 20 disabled campers comes to the tree house, where they get one-on-one attention by trained youth volunteers from all over the country, who also rotate weekly. Professional camp staff is on-hand to handle any behavioral anomalies, as well as direct activities and social events, Winter said.

In addition to the tree house, the camp has paved, wheelchair-friendly hiking trails, a petting zoo, volleyball, baseball, arts and crafts, and campfire sing-alongs.

Although the age of campers has ranged from 8 to 90 years old, the majority are adults in the late 30s to 50s, "with a handful of younger teens," Winter said.

Camp Matz was made possible through the assistance of Forever Young, and handicapped kids and adults of all ages around the country now have full access to tree houses designed especially for their needs, with long wheelchair ramps and other features that give them an opportunity to enjoy nature and social gatherings.

Forever Young has completed 16 tree houses for the handicapped from Vermont to California, and is planning on building more. The nonprofit organization is the brainchild of Bill Allen, an insurance agent from Burlington, Vermont, who was inspired to build accessible tree houses after reading a book by Seattle tree house builder Peter Nelson.

The organization also offers site evaluation and planning; tree surveys by a licensed arborist; architectural services; construction organization and management; project budgeting assistance; fund-raising; and tree house maintenance services.

Glossary

Arborist: a professional in the practice of arboriculture, which is the management and maintenance of ornamental or shade trees.

Balustrade: a vertical row of small posts (balusters) that support the upper beam of a railing.

Butt-and-cut: a method of creating a joint by placing one piece of wood on another at a right angle and cutting to the correct size and length.

Butt walls: the two walls in a room that provide support and that are overlapped by the through walls; butt-and-through walls stand in opposition to each other.

Cantilevering: a protruding horizontal beam fixed at a certain point.

Carabiner: an oblong or loop-shaped metal spring or screwed clip, usually attached to a rope.

Carriage bolt: wood bolts with a round head and square neck. The square neck under the round head sets into the hole in the wood to prevent turning and provide a secure fit.

Balustrade

Carabiner

Compartmentalization: the process trees use to heal injuries. The tree forms a barrier (compartment) around the damaged area, seals it off, and stops sending nutrients to the area. The tree then continues to grow around the damaged spot. By using a single, large bolt instead of several bolts placed closely together, the tree heals the injury over time, and both tree and support bolt remain strong, without infection.

Cripple stud: A cripple stud is a type of stud that is used above a door or above or beneath a window. They are not very different from other studs, except that they are cut shorter so there is room for the opening. The making of a cripple stud depends greatly on the size of the object that will be put around it.

Diameter: a straight-line segment that passes through the center of a circle and whose endpoints are on the circle.

Doorjamb: the vertical portion of the frame onto which a door is secured. Most types of door fasteners and deadbolts extend into a recess in the doorjamb when engaged, making the strength of the doorjambs vitally important to the overall security of the door.

Eyebolts: a screw with a loop on one end and threads on the other end. Eyebolts are commonly used to attach cables to objects.

Galvanized: steel-coated with a thin layer of zinc for rust and corrosion resistance.

Garnier Limb: a heavy bolt that embodies the universal idea of installing a "branch" where one is needed. GLs are a high-strength solution for supporting tree houses. They were developed and extensively experimented with by Michael Garnier, of the Out 'n' About Treesort in Takilma, Oregon.

Garnier Limb

Header: a structural member within the frame of a house, used as a support structure for windows, doors, and other openings.

Hip roof: a roof that slopes upward from all sides of a structure, without vertical ends.

Jack rafters: a rafter that meets a hip instead of the ridge board in a hip roof. It has an inclined top where it meets the hip rafter and is shorter than the common rafters.

Joist: the basic horizontal structural support for a ceiling, roof, or floor consisting of wood, steel, or concrete. Joists are smaller than beams and may run from wall-to-wall, wall-to-beam, or beam-to-beam. They are usually supported by beams.

Knee brace: a supporting structure for a tree house that consists of a level beam supported by braces attached to the tree, usually at a 45-degree angle.

Lag bolt

Lag bolt: a type of heavy-duty screw that, because of its size, appears to be a bolt.

Mason's line: a braided chord which, when saturated with chalk and secured between two coordinates, will "snap" an accurate, straight line.

Miter: an angled surface on a piece of wood formed to butt against an angled surface on another piece of wood so they can be joined.

Nailers: temporary wooden braces used to hold beams or other structural members in place during construction.

Mason's line

Pawl: a hinged device that articulates with the notch of a ratchet to provide forward motion and/or prevent backward motion.

Pilot hole: a small hole drilled into wood before placing a larger drill bit or screw into the wood. The pilot hole helps to prevent splitting of the wood and allows one to insert the larger screw straight.

Platform: a flat wooden structure that serves as the foundation for a tree house.

Pilot hole

Plumb: on a vertical plane, or up and down; as in when someone says, "level that wall," the correct carpentry term is "plumb that wall."

Purlin: a horizontal piece of wood supporting the rafters of a roof.

Rip cut: a method of cutting wood along its grain, as opposed to across the grain.

Rip cut

Rise-run ratio: a method of determining the slope or pitch of a roofline by comparing elevation to horizontal distance; an 8-in-12 roof rises 8 inches for every 12 inches of horizontal distance.

Square: a tool used to ensure that objects are perpendicular to one another.

Stringer: the structural member in a stairway that supports the treads and risers

Stud: an upright post in the framework of a wall for supporting sheets of lath, wallboard, or similar material.

Toe-nail: driving a nail at an angle through a board. Toe-nailing not only makes a strong joint, but also is a way to coax stubborn boards into position.

Square

Thimble: a terra cotta or metallic tube for the purpose of allowing a stovepipe to pass through a wooden construction without danger of fire.

Timber-frame construction: a method of building a structure, such as a tree house, without using nails, bolts, or other intrusive metal parts that penetrate the tree bark. The structure is held together with notched and pegged joints.

Viewshed: the natural environment visible from various vantage points. The term is used frequently in environmental law to describe areas to be protected from development.

Wall plate: a horizontal structural element at a right angle to vertical, weight-bearing supports, used for structural integrity in light-frame construction.

Bibliography

Anthenat. K.S., **American Tree Houses and Play Houses,** Betterway Publications, Crozet, VA, 1991.

Barkley, M., **Build Your Own Treehouse: A Practical Guide,** Sterling Publishing, New York, 2007.

Harris, J., **A Treehouse of Your Own,** Barron's, London, 2004.

Laurens, A., Dufour, D, André, G. **Treehouse Living,** Abrams, New York, 2007.

Lewis, S.P., Walsh, T.B.R., **Treehouse Chronicles,** TMC Books, Conway, N.H., 2005.

Trulove, J.G. **Tree Houses by Architects,** HarperDesign Int'l., New York, 2004.

Nelson, P. **Treehouses,** Houghton Mifflin, Boston, 1994.

Nelson, P. and J., **The Treehouse Book,** Universe Publishing, New York, 2000.

Stiles, D. and J., **Treehouses & Playhouses You Can Build,** Gibbs Smith, Salt Lake City, 2006.

Stiles, D. and J., **Tree Houses You Can Actually Build,** Houghton Mifflin, Boston, 1998.

The Complete Guide: Build Your Kids a Treehouse, Creative Publishing, Minneapolis, 2007.

Helpful Resources

Construction & Construction Tips

"Backyard Activities Building a Tree House or Fort" Free Online Library. February 14, 2009.
www.thefreelibrary.com/Backyard+Activities+Building+a+Tree+House+or+Fort-a01073932125

"Build Your Own Tree House" The Tree House Guide, 2009.
www.thetreehouseguide.com

"Building & Construction" Librarians' Internet Index, May 14, 2007.
www.lii.org/pub/subtopic/686

"Building a Treehouse" Home Improvement Articles, CornerHardware.com, 2007.
www.cornerhardware.com/home_improvement_articles/building_a_treehouse/61

"Build a Tree House for Your Kids" Servicemagic, 2009.
www.servicemagic.com/article.show.Build-a-Tree-House-for-Your-Kids.14780.html

"Build Your Own Tree House" Parent Central, 2008.

     www.parentcentral.ca/parent/article/543543

"Construction Images" Treehouse Chronicles.

     http://treehouseguy.wordpress.com/treehouse-construction-images

"Custom Tree Houses" Tree Top Builders, 2009.

     www.treetopbuilders.net

"Farrell Tree House Construction 2002" Bernard Farrell, 2002.

     www.bernardfarrell.com/thpics1.htm

"The GL Method" TreeHouse People.

     www.treehouse.jp/thp_eng/gl.html

"Here's How to Build a Treehouse" Farmer's Market Online, 2003.

     www.farmersmarketonline.com/howto20.htm

"How to Build a Treehouse" HubPages, 2009.

     http://hubpages.com/hub/How-to-Build-a-Treehouse

"How to Build a Treehouse" Ron Hazelton's HouseCalls.

     www.ronhazelton.com/archives/howto/treehouse_construction.shtm

"How to Build a Tree House From Pallets" eHow, 2009.

     www.ehow.com/how_4581231_build-tree-house-pallets.html

"How to Join The Treehouse Movement" InsideOut, August 2008.

     www.insideouthv.com/Earth/earth_treehousemovement.html

"How Tree Houses Work" How Stuff Works, 2009.

     http://home.howstuffworks.com/home-improvement/energy-efficiency/treehouse.htm

"How Will a Treehouse Affect Your Property?" Real Estate Pro Articles, December 10, 2008.

     www.realestateproarticles.com/Art/3362/264/How-Will-a-Treehouse-Affect-Your-Property.html

Living Tree Online. 2006. Living Tree, LLC.

     www.thelivingtreehouse.com

"Some Simple Bits for Building a Treehouse" Out'N'About Treehouse Construction. February 1998.

     www.treehouses.com/treehouse/construction/home.html

"Some Treehouse Construction Tips" Beverly Madera Outdoors.

www.beverlymadera.com/treehouse/constideas.html

"Tree House Building Companies" The Tree House Guide, 2009.

/www.thetreehouseguide.com/links-builders.htm

"Treehouse by Design Blog" Treehouse by Design. August 9, 2009.

www.treehousebydesign.com/blog

"Treehouse Construction Advice?" Ask Metafilter. September 4, 2007.

http://ask.metafilter.com/70714/Treehouse-construction-advice-photos-video

Treehouse Design & Construction. Northeast Treehouse.

www.northeasttreehouse.com

Treehouse Guides, 2009.

www.treehouseguides.com

TreeHouse Workshop. TreeHouse Workshop, Inc.

www.treehouseworkshop.com

"Up A Tree" Custom Home Online. November 1, 2007.

http://customhomeonline.name/industry-news.asp?sectionID=0&articleID=611435

Private Tree Houses

"10 Amazing Tree Houses from Around the World: Sustainable, Unique and Creative Designs" Web Urbanist, February 10, 2008.

http://weburbanist.com/2008/02/10/10-amazing-tree-houses-from-around-the-world-sustainable-unique-and-creative-designs/

"Dickinson College's "Treehouse" Captures Gold Rating for Environmental Design" College Campus News, July 21, 2008.

www.collegenews.org/x8725.xml

"The Derb Tree House" John Derbyshire, May 2004.

www.olimu.com/Notes/TreeHouse/TreeHouse.htm

"Erik's Treehouse on a Pine Tree."

    www.laading.net/erik/treehouse/pinetree.htm

"Fine Treehouse Building" iPoet.com, February 7, 2009.

    www.ipoet.com/Treehouses

"Home Sweet Treehouse" Mother Earth News, September 2001.

    www.motherearthnews.com/Nature-Community/2001-08-01/Home-Sweet-Treehouse.aspx?page=6

"The Lee Tree House" Architectural Record, October 2005.

    www.archrecord.construction.com/projects/residential/archives/0510treeHouse.asp

"O2 Sustainability Treehouse" Inhabitat, March 15, 2007.

    www.inhabitat.com/2007/03/15/o2-sustainability-treehouse-by-dustin-feider

"Our Treehouse" Shechter Family.

    www.shechterfamily.com/treehouse/index.html

"Steel Tree House" Architectural Record, 2009.

    http://archrecord.construction.com/projects/residential/archives/0506HotM-1.asp

"The Treehouse" Joe Petrie.

    www.joepetrie.org/treehouse.htm

"Tree Houses" Tréndir, May 14, 2009.

    www.trendir.com/house-design/tree_houses

"Tree-house Magic" Off-Grid, May 24, 2005.

    www.off-grid.net/2005/05/24/tree-house-magic

TreeHouses.org, 2008. Forever Young Treehouses.

    www.treehouses.org

"Tree House Page" SeeBee65, November 2007.

    www.matadorcoupe.com/seebee65/treehouse.htm

"Tree-Office" White Pine Consulting, May 30, 2007.

www.whitepinesconsulting.com/dandv/TreeOffice/index.html

"Tree Schooling: The Newest Branch of Homeschooling." Grand Oaks Acadamy Tree House.
www.grandoakstreehouse.homestead.com/Grandoakstreehouse.html

"The Ultimate Treehouse" "The Ultimate Treehouse" GeekDad, August 26, 2008.
www.wired.com/geekdad/2008/08/the-ultimate-tr

"The Walnut Treehouse by Baumraum" ConstructionBlog.org, August 29, 2008.
http://constructionblog.org/the-walnut-treehouse-by-baumraum

Safety & Education

"Climb Wave Treehouses Lure Children Outside." Go Erie.
www.saveourtreehouse.com/SaveOurTreehouse/links/frontyard_treehouse/frontyard_treehouse.htm

"New National Study Emphasizes Need For Tree House Safety Standards." Medical News Today, March 3, 2009.
www.medicalnewstoday.com/articles/140813.php

"Sustainable Treehouse Design and Construction" Yestermorrow Design/Build School.
www.yestermorrow.org/courses/wbc/treehouse.htm

"Treehouse Breaks The Mold, County Rules" The Seattle Times, May 27, 2008.
http://seattletimes.nwsource.com/html/localnews/2004440003_treehouse27m.html

"Treehouse Safety" Home Safety Zone, 2006.
www.homesafetyzone.com/backyard-safety/treehouse-safety/

Tree House Resorts/Get-Aways

"Tree House in Costa Rica Near Corcovado National Park" YouTube, January 14, 2009.
www.youtube.com/watch?v=w3YiocDXtjs

"Cedar Creek Treehouse at Mt. Rainer" Cedar Creek Treehouse.
www.cedarcreektreehouse.com

"Top 10 Treehouse Resorts" America's Best and Top Ten.

> **www.americasbestonline.net/treehouse.htm**

"Tree Houses – Escape From America And Live In A Tree House" EscapeArtist.com, 2004.

> **www.escapeartist.com/unique_lifestyles/Tree_Houses.html**

"Treehouse Restaurant Fit For An Elven Queen" TreeHugger, March 20, 2009.

> **www.treehugger.com/files/2009/03/treehouse_resta.php**

"The Tsala Treetop Lodge, An African Treehouse" Best House Design, July 17, 2008.

> **www.besthousedesign.com/2008/07/17/the-tsala-treetop-lodge-an-african-treehouse**

"Ultimate Treehouses at the Dallas Arboret" Douglas Newby & Associates, 2009.

> **www.dougnewby.com/architecture/treehouses.asp**

"Urban Tree House Programs" Natural Inquirer, 2008.

> **www.naturalinquirer.org/urban-treehouse-v-31.html**

"With a pirate ship, cave and tree house as offices, these designers may never come home" Post-Gazette, October 26, 2006.

> **www.post-gazette.com/pg/06299/733053-28.stm**

Author Biography

Robert M. Miskimon is the author of five published works of fiction, including *Skagit, Plastic Jesus,* and *What Death Can Touch.* He has written for the Associated Press, CBS/Merdscape, United Press International, and various daily and weekly newspapers. A native of Richmond, Virginia., he lives on Vashon Island, Washington.

Index